No Bullshit Guide to MATHEMATICS

Ivan Savov

January 5, 2021

No Bullshit Guide to Mathematics
by Ivan Savov

Published by Minireference Co.
Montréal, Québec, Canada
minireference.com | @minireference | fb.me/noBSguide

For inquiries, contact the author at ivan@minireference.com

Mathematics Subject Classifications (2010): 00A05, 00A06, 00A09, 97H20.

Fifth edition
v5.4 git commit 1268:5d39705

978-0-9920010-3-2

10 9 8 7 6 5 4 3 2 1

Contents

Concept maps

The diagrams below show all the topics and concepts covered in this book. Topics are shown in rectangular boxes. Concepts are shown in bubbles with rounded corners. Note the numerous connections.

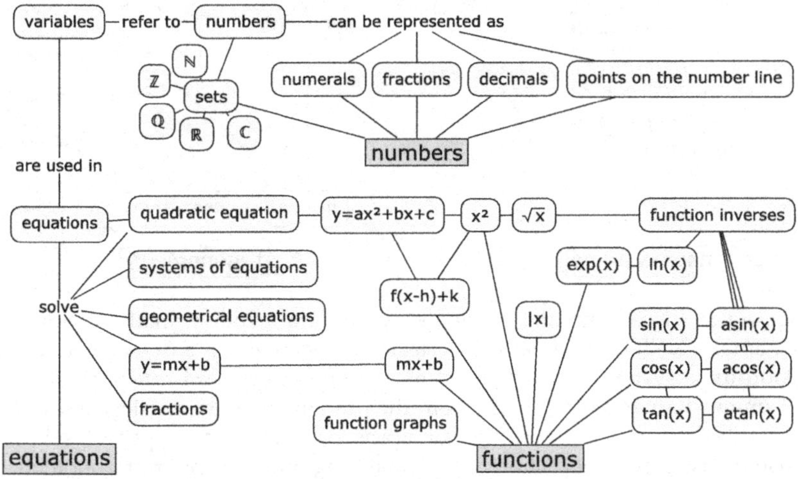

Figure 1: The first chapter of the book covers essential concepts like numbers, number representations, equations, and functions.

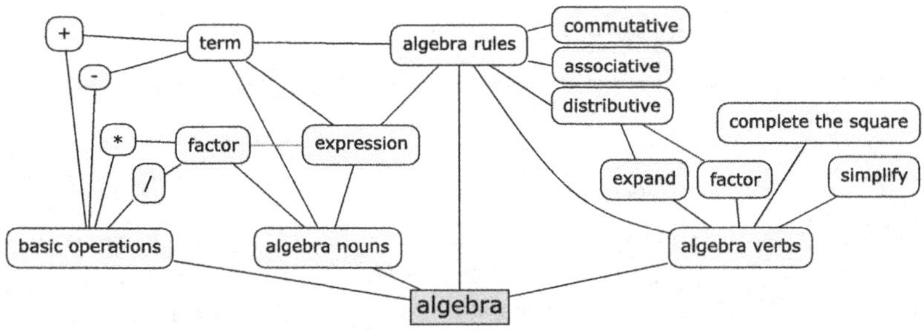

Figure 2: In Chapter 2 we'll learn about the rules of algebra, which codify what we're allowed to do with math expressions.

You can annotate the concept maps with your current knowledge of each concept to keep track of your progress. Add a single dot (•) next to all concepts you've heard of, two dots (••) next to concepts you think you know, and three dots (•••) next to concepts you've used in exercises and problems. By collecting some dots every week, you'll be able to move through the material in no time at all.

If you don't want to mark up your book, you can download a printable version of the concept maps here: `bit.ly/mathcmap`.

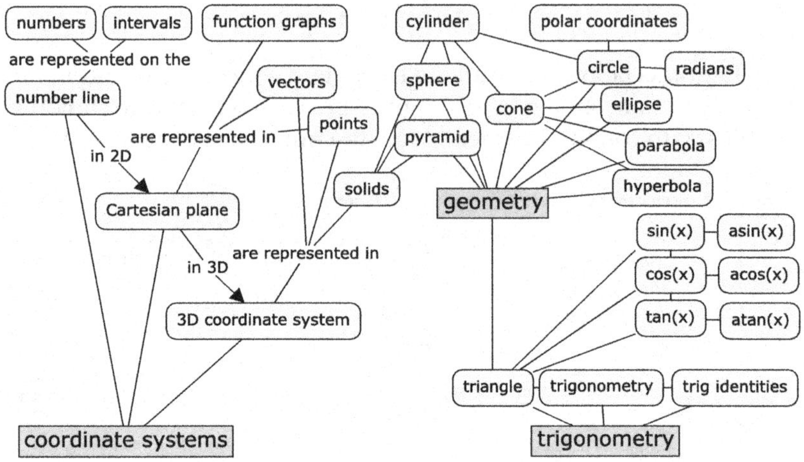

Figure 3: Visual thinking plays a very important role in mathematics. In Chapter 4 we'll learn about the Cartesian plane, which is a two-dimensional coordinate system used to represent points, vectors, and function graphs. Then in Chapter 6 we'll learn about the fundamental geometric shapes like triangles, pyramids, cones, circles, ellipses, parabolas, and hyperbolas. We'll also dedicate several sections to familiarizing ourselves with trigonometry, trigonometric functions, and trigonometric identities.

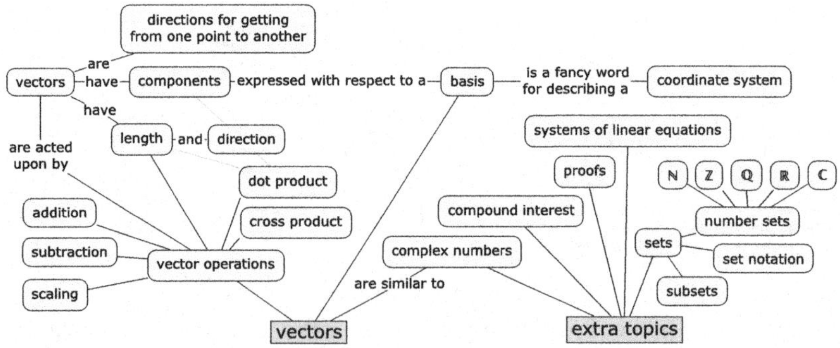

Figure 4: In Chapter 7, we'll use our algebra and geometry skills to learn about vectors and vector operations. In Chapter 8, we'll discuss math applications and formal math methods like proofs.

Consult the index on page 231 to find the exact location in the book where each concept is defined.

Preface

Why this book?

Math is a beautiful subject with many practical applications. Yet many people feel uncomfortable around math. Some even claim to "hate" math and try to avoid the subject. Countless adults are living with *math phobia*. That's too bad because learning math opens many doors to understanding concepts in science, computing, music, and business. Math is also useful for accomplishing day-to-day tasks.

Why learn math?

Math is a universal language for describing patterns and shapes. Math is everywhere around us. When you learn math, you'll be able to see the equations behind real-world phenomena, and then use these equations to predict the future. How cool is that? Learning math will also help you make better daily decisions like avoiding predatory loans, estimating times and distances when travelling, double-checking calculations in contracts and bills, and forecasting the cash flow of your business. Math is not just for scientists and engineers—*everyone* benefits from knowing math the same way everyone benefits from knowing how to read and write.

Is this book for you?

This book is written for adult learners who think they suck at math and feel uncomfortable around numbers and equations. Whether you hate math because of bad experiences, feel disheartened at the sight of equations, or simply don't see the point of learning math, this book has something to offer you. It's time to reconnect with this beautiful subject.

This book will be useful for everyone who wants to learn math. High school students can quickly learn the material they need for their classes. University students can use the book to review and prepare for more advanced subjects like calculus, statistics, and linear algebra. Parents who want to help their kids with homework can depend on this book to refresh their math skills.

How are we going to do this?

This book approaches math by showing you the connections between concepts. Learning math does not require memorizing facts and equations. It's not about how many equations you know, but about knowing how to get from one equation to another. Consult the concept maps on pages iii and iv to see all the topics covered in the book and how they are interconnected.

Each section in this book is a short tutorial that covers one math topic with precise definitions, illustrative examples, and concise explanations. By progressing through the chapters and sections of this book, you'll develop your math knowledge step by step, with each new layer of understanding building upon the previous ones. The book covers numbers, sets, variables, equations, algebra, functions, geometry, trigonometry, and also math applications to personal finance. If it's useful math, we've got it covered! To ensure that you *really* grasp the material, the book provides numerous exercises and practice problems (see Chapter 9). You'll find the answers to all the exercises and problems in Appendix A (page 205).

About the publisher

Minireference Co. is a publisher specializing in math and science. Our goal is to make advanced math modelling tools accessible to everyone by producing affordable textbooks that explain math concepts clearly and concisely. It's time for a break from traditional textbooks that are expensive, heavy, and tedious to read. The books in the **No Bullshit Guide** series are an efficient and enjoyable alternative for learning mathematical subjects. Whether you're a student, a parent, or an adult learner, we've got a book for you that will help you level-up in math.

About the author

I have been tutoring math and physics for more than 17 years. Through this experience, I learned to break complicated ideas into smaller, interconnected chunks that are easy to understand. I think the best way to teach math is to clearly define concepts and show the paths that connect them. When learners know the basic vocabulary and are able to find their way in a new mathematical domain, the battle against *math phobia* is already half won.

Ivan Savov
Montreal, 2020

Introduction

If you look carefully, you'll find mathematical patterns all around you: complex patterns like the water waves on the surface of a lake, medium-complexity patterns like the up-and-down oscillations of a mass suspended on a spring, and basic patterns that occur in all numbers and geometric shapes.

This is a book about basic math patterns. The goal is to introduce you to fundamental building blocks like numbers, expressions, functions, algebra, and geometry. Understanding basic math concepts will provide you with a common core of tools that you can use in diverse fields like physics, chemistry, biology, medicine, engineering, and business.

Thinking like a mathematician means being able to notice the patterns everywhere around you. That's what mathematicians do all day—seriously! First you *observe* a pattern; then you try to *describe* the pattern (this is usually where the equations and formulas come in); finally you *use* the pattern to achieve some practical outcome. Would you believe me if I told you that a few equations scribbled on a paper napkin can predict the outcome of a quantum physics experiment, calculate the yield of a chemical reaction, predict the growth rate of bacteria, save a patient's life, build a bridge, and maximize the profits of a startup?

Do I have your attention? I hope at this point you're interested in learning at least a little bit of math. Or perhaps you still feel kind of weird about reading a math book, and extra weird about reading a math book by choice, without someone forcing you to read it. What kind of trouble are you getting yourself into?

Relax. Math is no trouble. Learning math is not something to worry about or foresee as difficult. Math is not some sort of obstacle you have to jump over in one shot. Learning math is more like climbing a mountain. It's up to you to choose the pace of your climb. By investing some time in reading and solving exercises, everyone can gradually make it up the mountain.

The trick to learning math is to redefine what math means to you.

Dear reader, trust me on this one: math does not involve memorization and rote calculations! That's arithmetic. It's great if you're fast at adding, subtracting, multiplying, and dividing, but real math is so much more than that. For example, learning some basic algebra rules will allow you to simplify complicated expressions by rewriting them as equivalent but shorter expressions, so by the time you start plugging in the numbers you have a lot less arithmetic to do. The more algebra you learn, the less arithmetic you'll need to do. What? We learn more math in order to do less math? Yes, that's pretty much the whole point of the subject. Knowing how to recognize abstract patterns, how to describe them precisely, and how to use them to simplify problems turns out to be a very useful skill set.

In addition to being a useful tool for learning other sciences, learning math is interesting on its own. There will be some beautiful views at the end of each climb up the math mountains.

Guide to the math mountains

This book is your guide for the first part of your journey in the math mountains. Math concepts like numbers, equations, functions, algebra, and geometry are the first peaks you must climb on your way to the major math and science mountains.

In Chapter 1, we'll start by defining fundamental math ideas including numbers, variables, equations, and functions. These are the main building blocks of math, and we'll use them throughout the book. Readers who are new to mathematics should read Chapter 1 carefully and invest the time needed to solve the exercises.

Algebra! Yes, there, I said it. In Chapter 2 we'll learn the rules of algebra, which consist of various simplifications, reductions, and expansions we can perform on math expressions. Learning algebra by memorizing a bunch of rules is a very difficult task. Instead, you can think of algebra as a language. Algebra techniques are *verbs* that we apply to math expressions. To expand the expression $a(b + c)$ means to write the expression as $ab + ac$. To factor the expression $ax + ay + az$ is to write it as $a(x + y + z)$. To learn algebra you'll need to learn some new nouns: *variable, term, factor, expression, equation*. You'll also need to become familiar with the algebra verbs: *to solve* (equations), *to factor, to simplify, to expand* (expressions), *to complete the square* (in a quadratic expression), etc.

Are you still with me, dear reader? I know the prospects of learning algebra might not seem like the most exciting thing ever, and I'm not going to lie to you and say it will be easy to develop your expression-simplifying skills, but the payoff is totally worth it.

In Chapter 3 we'll discuss exponents and logarithms in detail.

For instance, the expression a^x means to multiply a by itself x times. Exponents have some interesting properties, which we'll spend some time exploring. We'll also discuss the logarithmic function $\log_a(x)$, which is the inverse function of the exponential function a^x.

In Chapter 4 we'll discuss the Cartesian plane, which allows us to visualize math concepts like points, vectors, and geometric shapes. We can also use the Cartesian plane to plot function graphs, which are visual representations of how functions behave for all inputs.

Chapter 5 is dedicated to the study of functions. We'll introduce important new vocabulary for describing functions' inputs, outputs, and other properties. We'll build a catalogue of the 10 most important functions for math and scientific modelling. We'll also discuss how to combine, modify, and transform basic functions in order to form composite functions with various desirable properties.

The subject of geometry deals with shapes, transformations, distances, areas, and volumes. In Chapter 6 we'll learn about geometry, and discuss various formulas for lines, circles, cylinders, spheres, and other shapes. We'll also discuss trigonometry—the study of the relative lengths of the sides of right-angle triangles.

In Chapter 7 we'll discuss vectors, which are used to describe directions in space. Vectors are useful for science calculations, computing, statistics, machine learning, and myriad other domains.

In Chapter 8 we'll discuss various supplementary math topics. We'll see how to solve systems of linear equations, how to compute compound interest, and learn a bit about proofs and set notation. Learning set notation (Section 8.3) is a key foundation for studying more advanced math topics.

The book concludes with a sizeable set of problems (Chapter 9). Take note, dear reader: math is not a spectator sport. Reading math is not enough—you must also practice your skills by solving problems.

Use the maps

The concept maps on pages iii and iv show an overview of the concepts and topics we'll discuss in the book. Note the numerous connections between concepts. Understanding these connections is the key to understanding math: math isn't a random bunch of facts, but an interconnected network of ideas.

Are you ready for this? Let's get started!

Chapter 1

Numbers, variables, and equations

In this chapter we'll introduce the fundamental building blocks of math: numbers, variables, equations, and functions. Learning about the basics will establish a solid foundation for all other topics that are to come.

We'll start by learning a general procedure for solving equations. This is a fundamental skill you'll develop throughout the book, starting in Section 1.1 and expanding as you learn about more advanced types of equations and solution techniques. In Section 1.2 we'll define the different types of numbers that exist, then discuss number representations in Section 1.3. In Section 1.4 we'll learn how to use variables like k and x to represent unknowns in equations.

We'll then discuss functions and function inverses in Section 1.5. A function describes a transformation between an input variable x and a corresponding output variable y. The inverse function describes the same transformation in the opposite direction, treating y as the input and x as the output. By learning how to apply functions and function inverses to solve equations, you're taking one giant step toward acquiring math superpowers. Overall, your knowledge of numbers, variables, equations, and functions will allow you to solve many types of problems on the journey ahead.

1.1 Solving equations

Most math skills boil down to being able to manipulate and solve equations. Solving an equation means finding the value of the unknown in the equation.

Check this shit out:

$$x^2 - 4 = 45.$$

To solve the above equation is to answer the question "What is x?" More precisely, we want to find the number that can take the place of x in the equation so that the equality holds. In other words, we're asking,

"Which number times itself minus four gives 45?"

That is quite a mouthful, don't you think? To remedy this verbosity, mathematicians often use specialized symbols to describe math operations. The problem is that these specialized symbols can be very confusing. Sometimes even the simplest math concepts are inaccessible if you don't know what the symbols mean.

What are your feelings about math, dear reader? Are you afraid of it? Do you have anxiety attacks because you think it will be too difficult for you? Chill! Relax, my brothers and sisters. There's nothing to it. Nobody can magically guess the solution to an equation immediately. To find the solution, you must break the problem into simpler steps. Let's walk through this one together.

To find x, we can manipulate the original equation, transforming it into a different equation (as true as the first) that looks like this:

$$x = \text{only numbers}.$$

That's what it means to *solve* an equation: the equation is solved because the unknown is isolated on one side, while the constants are grouped on the other side. You can type the numbers on the right-hand side into a calculator and obtain the numerical value of x.

By the way, before we continue our discussion, let it be noted: the equality symbol ($=$) means that all that is to the left of $=$ is equal to all that is to the right of $=$. To keep this equality statement true, **for every change you apply to the left side of the equation, you must apply the same change to the right side of the equation**.

To find x, we need to manipulate the original equation into its final form, simplifying it step by step until it can't be simplified any further. The only requirement is that the manipulations we make transform one true equation into another true equation. In this example, the first simplifying step is to add the number four to both sides of the equation:

$$x^2 - 4 + 4 = 45 + 4,$$

which simplifies to

$$x^2 = 49.$$

Now the expression looks simpler, yes? How did I know to perform this operation? I wanted to "undo" the effects of the operation -4. We undo an operation by applying its *inverse*. In the case where the operation is the subtraction of some amount, the inverse operation is the addition of the same amount. We'll learn more about function inverses in Section 1.5.

We're getting closer to our goal of *isolating* x on one side of the equation, leaving only numbers on the other side. The next step is to undo the square x^2 operation. The inverse operation of squaring a number x^2 is to take its square root $\sqrt{\ }$, so that's what we'll do next. We obtain

$$\sqrt{x^2} = \sqrt{49}.$$

Notice how we applied the square root to both sides of the equation? If we don't apply the same operation to both sides, we'll break the equality!

The equation $\sqrt{x^2} = \sqrt{49}$ simplifies to

$$|x| = 7.$$

What's up with the vertical bars around x? The notation $|x|$ stands for the *absolute value* of x, which is the same as x except we ignore the sign that indicates whether x is positive or negative. For example $|5| = 5$ and $|-5| = 5$, too. The equation $|x| = 7$ indicates that both $x = 7$ and $x = -7$ satisfy the equation $x^2 = 49$. Seven squared is 49, $7^2 = 49$, and negative seven squared is also 49, $(-7)^2 = 49$, because the two negative signs cancel each other out.

The final solutions to the equation $x^2 - 4 = 45$ are

$$x = 7 \qquad \text{and} \qquad x = -7.$$

Yes, there are *two* possible answers. You can check that both of the above values of x satisfy the initial equation $x^2 - 4 = 45$.

If you are comfortable with all the notions of high school math and you feel you could have solved the equation $x^2 - 4 = 45$ on your own, then you can skim through this chapter quickly. If on the other hand you are wondering how the squiggle killed the power two, then this chapter is for you! In the following sections we will review all the essential concepts from high school math that you will need to power through the rest of this book. First, let me tell you about the different kinds of numbers.

1.2 Numbers

In the beginning, we must define the main players in the world of math: numbers.

Definitions

Numbers are the basic objects we use to count, measure, quantify, and calculate things. Mathematicians like to classify the different kinds of number-like objects into categories called *sets*:

- The natural numbers: $\mathbb{N} = \{0, 1, 2, 3, 4, 5, 6, 7, \ldots\}$
- The integers: $\mathbb{Z} = \{\ldots, -3, -2, -1, 0, 1, 2, 3, \ldots\}$
- The rational numbers: $\mathbb{Q} = \{\frac{5}{3}, \frac{22}{7}, 1.5, 0.125, -7, \ldots\}$
- The real numbers: $\mathbb{R} = \{-1, 0, 1, \sqrt{2}, e, \pi, \ 4.94\ldots, \ \ldots\}$
- The complex numbers: $\mathbb{C} = \{-1, 0, 1, i, 1+i, 2+3i, \ldots\}$

These categories of numbers should be somewhat familiar to you. Think of them as neat classification labels for everything that you would normally call a number. Each group in the above list is a *set*. A set is a collection of items of the same kind. Each collection has a name and a precise definition for which items belong in that collection. Note also that each of the sets in the list contains all the sets above it, as illustrated in Figure 1.1. For now, we don't need to go into the details of sets and set notation, but we do need to be aware of the different sets of numbers.

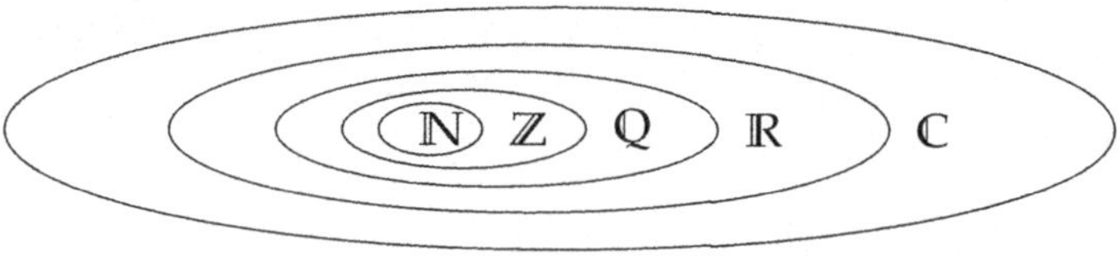

Figure 1.1: An illustration of the nested containment structure of the different number sets. The set of natural numbers is contained in the set of integers, which in turn is contained in the set of rational numbers. The set of rational numbers is contained in the set of real numbers, which is contained in the set of complex numbers.

Why do we need so many different sets of numbers? Each set of numbers is associated with more and more advanced mathematical problems.

The simplest numbers are the natural numbers $\mathbb{N}$, which are sufficient for all your math needs if all you're going to do is *count* things. How many goats? Five goats here and six goats there so the total is

11 goats. The sum of any two natural numbers is also a natural number.

As soon as you start using *subtraction* (the inverse operation of addition), you start running into negative numbers, which are numbers outside the set of natural numbers. If the only mathematical operations you will ever use are *addition* and *subtraction*, then the set of integers $\mathbb{Z} = \{\ldots, -2, -1, 0, 1, 2, \ldots\}$ will be sufficient. Think about it. Any integer plus or minus any other integer is still an integer.

You can do a lot of interesting math with integers. There is an entire field in math called *number theory* that deals with integers. However, to restrict yourself solely to integers is somewhat limiting—a rotisserie menu that offers $\frac{1}{2}$ of a chicken would be totally confusing.

If you want to use division in your mathematical calculations, you'll need the rationals $\mathbb{Q}$. The set of rational numbers corresponds to all numbers that can be expressed as *fractions* of the form $\frac{m}{n}$ where m and n are integers, and $n \neq 0$. You can add, subtract, multiply, and divide rational numbers, and the result will always be a rational number. However, even the rationals are not enough for all of math!

In geometry, we can obtain *irrational* quantities like $\sqrt{2}$ (the diagonal of a square with side 1) and π (the ratio between a circle's circumference and its diameter). There are no integers x and y such that $\sqrt{2} = \frac{x}{y}$, therefore we say that $\sqrt{2}$ is *irrational* (not in the set $\mathbb{Q}$). An irrational number has an infinitely long decimal expansion that doesn't repeat. For example, $\pi = 3.141592653589793\ldots$ where the dots indicate that the decimal expansion of π continues all the way to infinity.

Combining the irrational numbers with the rationals gives us all the useful numbers, which we call the set of real numbers $\mathbb{R}$. The set $\mathbb{R}$ contains the integers, the rational numbers $\mathbb{Q}$, as well as irrational numbers like $\sqrt{2} = 1.4142135\ldots$. By using the reals you can compute pretty much anything you want. From here on in the text, when I say *number*, I mean an element of the set of real numbers $\mathbb{R}$.

The only thing you can't do with the reals is to take the square root of a negative number—you need the complex numbers $\mathbb{C}$ for that. We defer the discussion on $\mathbb{C}$ until the end of Chapter 7.

Operations on numbers

Addition

You can add numbers. I'll assume you're familiar with this stuff:

$$2 + 3 = 5, \quad 45 + 56 = 101, \quad 9\,999 + 1 = 10\,000.$$

You can visualize numbers as sticks of different length. Adding numbers is like adding sticks together: the resulting stick has a length equal to the sum of the lengths of the constituent sticks, as illustrated in Figure 1.2.

Figure 1.2: The addition of numbers corresponds to adding lengths.

Addition is *commutative*, which means that $a + b = b + a$. In other words, the order of the numbers in a summation doesn't matter. It is also *associative*, which means that if you have a long summation like $a + b + c$ you can compute it in any order $(a + b) + c$ or $a + (b + c)$, and you'll get the same answer.

Subtraction

Subtraction is the inverse operation of addition:

$$2 - 3 = -1, \quad 45 - 56 = -11, \quad 999 - 1 = 998.$$

Unlike addition, subtraction is not a commutative operation. The expression $a - b$ is not equal to the expression $b - a$, or written mathematically:

$$a - b \neq b - a.$$

Instead we have $b - a = -(a - b)$, which shows that changing the order of a and b in the expression changes its sign.

Subtraction is not associative either:

$$(a - b) - c \neq a - (b - c).$$

For example $(7 - 2) - 3 = 2$ while $7 - (2 - 3) = 8$.

Multiplication

You can also multiply numbers together:

$$ab = \underbrace{a + a + \cdots + a}_{b \text{ times}} = \underbrace{b + b + \cdots + b}_{a \text{ times}}.$$

Note that multiplication can be defined in terms of repeated addition.

The visual way to think about multiplication is as an area calculation. The area of a rectangle of width a and height b is equal to ab. A rectangle with a height equal to its width is a square, and this is why we call $aa = a^2$ "a squared."

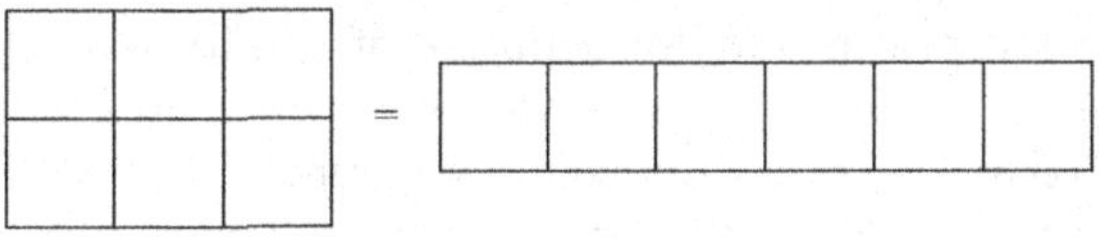

Figure 1.3: The area of a rectangle with width 3 m and height 2 m is equal to 6 m^2, which is equivalent to six squares with area 1 m^2 each.

Multiplication of numbers is also commutative, $ab = ba$, and associative, $abc = (ab)c = a(bc)$. In modern math notation, no special symbol is required to denote multiplication; we simply put the two factors next to each other and say the multiplication is *implicit*. Some other ways to denote multiplication are $a \cdot b$, $a \times b$, and, on computer systems, $a * b$.

Division

Division is the inverse operation of multiplication.

$$a/b = \frac{a}{b} = a \div b = \text{ one } b^{\text{th}} \text{ of } a.$$

Whatever a is, you need to divide it into b equal parts and take one such part.

Division is not a commutative operation since a/b is not equal to b/a. Division is not associative either: $(a \div b) \div c \neq a \div (b \div c)$. For example, when $a = 6$, $b = 3$, and $c = 2$, we get $(6/3)/2 = 1$ while $6/(3/2) = 4$.

Note that you cannot divide by 0. Try it on your calculator or computer. It will say "`error divide by zero`" because this action simply doesn't make sense. After all, what would it mean to divide something into zero equal parts?

Exponentiation

The act of multiplying a number by itself many times is called *exponentiation*. We denote "a exponent n" using a superscript, where n is the number of times the base a is multiplied by itself:

$$a^n = \underbrace{aaa \cdots a}_{n \text{ times}}.$$

In words, we say "a raised to the power of n."

To visualize how exponents work, we can draw a connection between the value of exponents and the dimensions of geometric objects. Figure 1.4 illustrates how the same length 2 corresponds to different geometric objects when raised to different exponents. The number 2 corresponds to a line segment of length two, which is a geometric object in a one-dimensional space. If we add a line segment of length two in a second dimension, we obtain a square with area 2^2 in a two-dimensional space. Adding a third dimension, we obtain a cube with volume 2^3 in a three-dimensional space. Indeed, raising a base a to the exponent 2 is commonly called "a squared," and raising a to the power of 3 is called "a cubed."

The geometrical analogy about one-dimensional quantities as lengths, two-dimensional quantities as areas, and three-dimensional quantities as volumes is good to keep in mind.

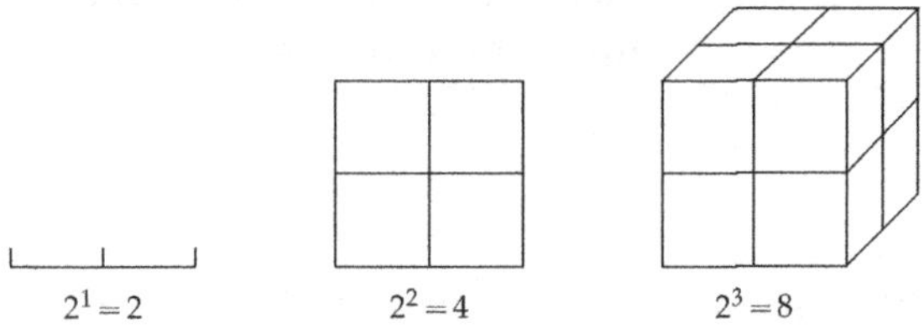

Figure 1.4: Geometric interpretation for exponents 1, 2, and 3. A length raised to exponent 2 corresponds to the area of a square. The same length raised to exponent 3 corresponds to the volume of a cube.

Our visual intuition works very well up to three dimensions, but we can use other means of visualizing higher exponents, as demonstrated in Figure 1.5.

Operator precedence

There is a standard convention for the order in which mathematical operations must be performed. The basic algebra operations have the following precedence:

1. Parentheses
2. Exponents
3. Multiplication and Division
4. Addition and Subtraction

If you're seeing this list for the first time, the acronym PEMDAS and the associated mnemonic "Please Excuse My Dear Aunt Sally," might help you remember the order of operations.

For instance, the expression $5 \cdot 3^2 + 13$ is interpreted as "First find the square of 3, then multiply it by 5, and then add 13." Parentheses are needed to carry out the operations in a different order: to multiply 5 times 3 first and *then* take the square, the equation should read

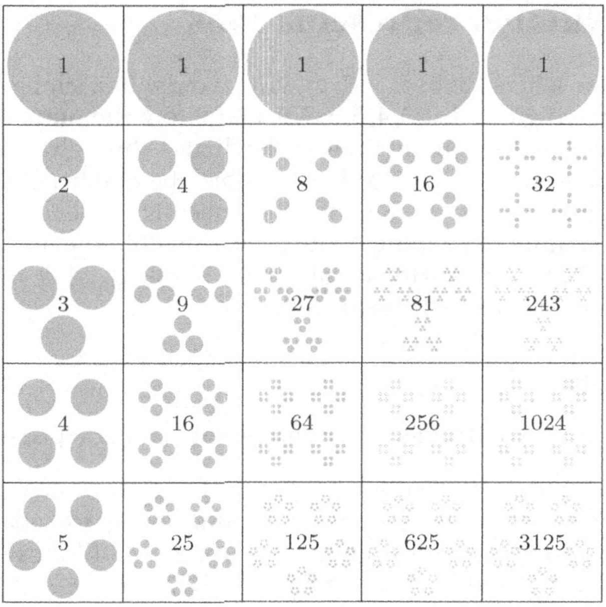

Figure 1.5: Visualization of numbers raised to different exponents. Each box in this grid contains a^n dots, where the base a varies from one through five, and the exponent n varies from one through five. In the first row we see that the number $a = 1$ raised to any exponent is equal to itself. The second row corresponds to the base $a = 2$ so the number of dots doubles each time we increase the exponent by one. Starting from $2^1 = 2$ in the first column, we end up with $2^5 = 32$ in the last column. The rest of the rows show how exponentiation works for different bases.

$(5 \cdot 3)^2 + 13$, where parentheses indicate that the square acts on $(5 \cdot 3)$ as a whole and not on 3 alone.

Exercises

E1.1 Solve for the unknown x in the following equations:

a) $3x + 2 - 5 = 4 + 2$ **b)** $\frac{1}{2}x - 3 = \sqrt{3} + 12 - \sqrt{3}$

c) $\frac{7x-4}{2} + 1 = 8 - 2$ **d)** $5x - 2 + 3 = 3x - 5$

E1.2 Indicate all the number sets the following numbers belong to.

a) -2 **b)** $\sqrt{-3}$ **c)** $8 \div 4$ **d)** $\frac{5}{3}$ **e)** $\frac{\pi}{2}$

E1.3 Calculate the values of the following expressions:

a) $2^3 3 - 3$ **b)** $2^3(3 - 3)$ **c)** $\frac{4-2}{3^3}(6 \cdot 7 - 41)$

1.3 Number representations

We use the letters "a, b, c, ..." to write words. In a similar fashion, we use the digits $0, 1, 2, 3, 4, 5, 6, 7, 8$, and 9 to write numbers in the language of math. You can think of the digits 0 through 9 as the "letters" used to write numbers. For example, the number 334 consists of the digits 3, 3, and 4. Note that the same digit 3 denotes two different quantities depending on its position within the number. The first digit 3 corresponds to the value three hundred, while the second digit 3 corresponds to the value thirty.

Concepts

In this section, we'll review three important number representations:

- The *decimal notation* for integers, rationals, and real numbers consists of an integer part and a fractional part separated by a *decimal point*. For example, the decimal 32.17 consists of the integer 32 and the fractional part 0.17.
- The *fraction notation* for integers and rational numbers consists of a numerator divided by a denominator. Here are some sample math expressions with fractions: $\frac{1}{2}$, $\frac{3}{4}$, $\frac{3}{2} = 1\frac{1}{2}$, and $\frac{17}{100}$.
- The *number line* is a graphical representation for numbers that allows us to visualize numbers as geometric points on a line.

The same number a can be represented in multiple equivalent ways. It is often convenient to convert from one representation to another depending on the calculations we need to perform. For example, the number three can be expressed as the numeral 3, the decimal 3.0, the fraction $\frac{3}{1}$, or as the point that lies three units to the right of the origin on the number line. All these representations refer to the same quantity, but each representation is useful in different contexts. The goal of this section is to get you comfortable working with all the number representations.

Positional notation for numbers

The Hindu–Arabic numeral system is the most widely used system for writing numbers today. It is a *decimal positional* system. The term *decimal* refers to the fact that it uses 10 unique symbols (the digits 0 through 9) to represent numbers. The system is *positional* because the value of each digit depends on its position within the number. Positional number systems are also called *place-value* systems.

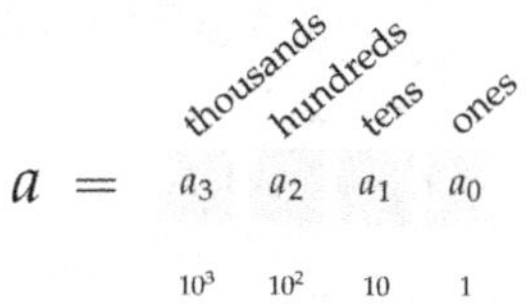

Figure 1.6: The place-value representation of the number $a = a_3a_2a_1a_0$.

Note the terminology used to refer to the individual digits of the numeral: we call a_3 the thousands, a_2 the hundreds, a_1 the tens, and a_0 the units.

Any natural number $a \in \mathbb{N}$, no matter how large, can be written as a sequence of digits:

$$\begin{aligned} a &= a_n \cdots a_2a_1a_0 \\ &= a_n \cdot 10^n + \cdots + a_2 \cdot 10^2 + a_1 \cdot 10 + a_0 \cdot 1, \end{aligned}$$

where the digits $a_0, a_1, \ldots$ come from the set $\{0,1,2,3,4,5,6,7,8,9\}$.

For example, the numeral 4235 corresponds to this calculation:

$$\begin{aligned} 4235 &= 4 \cdot 10^3 \;\; + 2 \cdot 10^2 + 3 \cdot 10 + 5 \cdot 1 \\ &= 4 \cdot 1000 + 2 \cdot 100 + 3 \cdot 10 + 5 \cdot 1 \\ &= 4000 + 200 + 30 + 5. \end{aligned}$$

Note how the English pronunciation of the number, "four thousand, two hundred and thirty-five," literally walks you through the calculation.

Decimal representation

Any number a less than one can be written as a *decimal point* followed by a sequence of digits, as illustrated in Figure 1.7.

$$\begin{aligned} a &= 0.a_{-1}a_{-2}a_{-3}\cdots \\ &= 0 + \frac{a_{-1}}{10^1} + \frac{a_{-2}}{10^2} + \frac{a_{-3}}{10^3} + \cdots. \end{aligned}$$

The decimal point indicates the beginning of the fractional part of a number. The place values of the digits to the right of the decimal point correspond to different decimal fractions. For example, the digit 7 corresponds to three different decimal fractions depending on its position within the number:

$$0.7 = \frac{7}{10}, \qquad 0.07 = \frac{7}{100}, \qquad \text{and} \qquad 0.007 = \frac{7}{1000}.$$

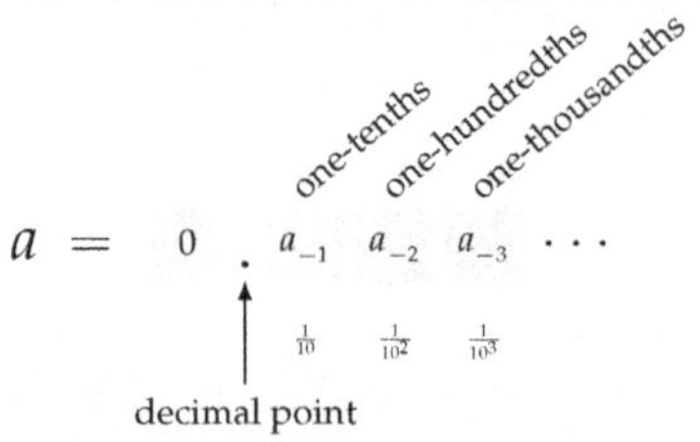

Figure 1.7: The decimal representation of a number smaller than one.

The first digit to the right of the decimal point a_{-1} represents the *tenths*, the second digit a_{-2} represents the *hundredths*, the third the *thousandths*, and so on.

In general, a number written in decimal notation has both an integer part and a fractional part:

$$\begin{aligned} a &= a_n \cdots a_2 a_1 a_0 \,.\, a_{-1} a_{-2} a_{-3} \cdots \\ &= a_n \cdot 10^n + \cdots + a_2 \cdot 10^2 + a_1 \cdot 10 + a_0 + \frac{a_{-1}}{10^1} + \frac{a_{-2}}{10^2} + \frac{a_{-3}}{10^3} + \cdots \end{aligned}$$

The decimal point appears in the middle of the digits and acts as a separator. The digits to the left of the decimal point, $a_n \cdots a_2 a_1 a_0$, correspond to the integer part of the number, while the digits to the right of the decimal, $0.a_{-1}a_{-2}a_{-3}\cdots$, correspond to the fractional part of the number.

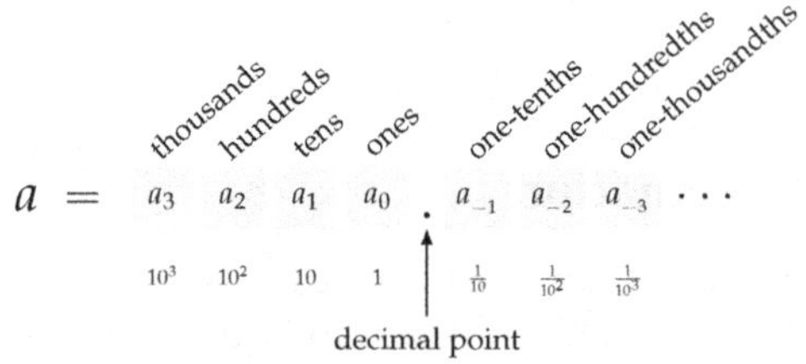

Figure 1.8: The decimal number a consists of an integer part $a_3a_2a_1a_0$ and a fractional part $0.a_{-1}a_{-2}\cdots$ separated by the decimal point.

Note the names for the different digits in the fractional part of the decimal in Figure 1.8. These names are used when we describe the fractional part of a decimal in words:

- "1.4" is read "one and four tenths," or you could informally describe the decimal as you see it written: "one point four."
- "45.37" is read "forty-five and thirty-seven hundredths," or sometimes "forty-five point three seven."
- A length measurement like "0.345 in" is read "three-hundred forty-five thousandths of an inch."

We can use decimal notation to represent rational numbers like one-half (0.5), one-quarter (0.25), and three-quarters (0.75). We can also use decimal notation to write approximations to irrational numbers. For example, the irrational number $\sqrt{2}$ (the diagonal of a square with length one) is approximately equal to 1.41421. We say the approximation 1.41421 is "accurate to five decimals," because this is how many digits there are in its fractional part.

* * *

So far we reviewed the decimal representation for numbers, which is very familiar to us from everyday life. Perhaps you're starting to think that math isn't so bad after all? Some of you must be saying, "Wonderful, I'm becoming friends with numbers while avoiding uncomfortable topics like fractions." Sorry, but you're not getting off so easily because this is exactly what's coming up next. That's right, we're about to make friends with fractions, too.

Fractions

Fractions describe what happens when a *whole* is cut into n equal parts and we are given m of those parts. For example, the fraction $\frac{3}{8}$ describes having three parts out of a whole cut into eight parts, hence the name "three-eighths."

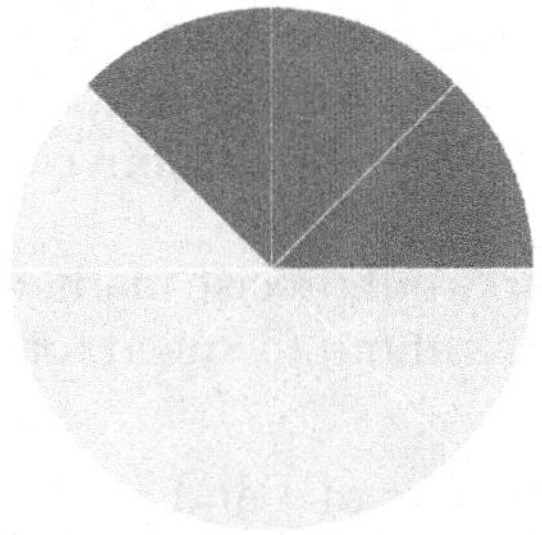

Figure 1.9: The fraction $\frac{3}{8}$ can be visualized as three slices from a pizza that has been cut into eight equal slices.

It's important to understand fractions because many math concepts like rational numbers ($\mathbb{Q}$), ratios, percents, and probabilities are best described in the language of fractions.

Definitions

The fraction "a over b" can be written three different ways:

$$a/b = a \div b = \frac{a}{b}.$$

The top and bottom parts of the fraction $\frac{a}{b}$ have special names:

- b is called the *denominator* of the fraction. It tells us how many parts make up the whole.
- a is called the *numerator*. It tells us the number of parts we have.

Fractions are the most natural way to represent rational numbers. Why natural? Check out these simple fractions:

$$\begin{aligned}
\tfrac{1}{1} &= 1.0 \\
\tfrac{1}{2} &= 0.5 \\
\tfrac{1}{3} &= 0.33333\ldots = 0.\overline{3} \\
\tfrac{1}{4} &= 0.25 \\
\tfrac{1}{5} &= 0.2 \\
\tfrac{1}{6} &= 0.166666\ldots = 0.1\overline{6} \\
\tfrac{1}{7} &= 0.142857142857142 85\ldots = 0.\overline{142857}
\end{aligned}$$

The fractional notation on the left is preferable because it shows the underlying *structure* of the number while avoiding the need to write complicated decimals.

When written as decimal numbers, certain fractions have infinitely long decimal expansions. We use the overline notation to indicate the digit(s) that repeat infinitely in the decimal, as in the case of $0.\overline{3}$, $0.1\overline{6}$, and $0.\overline{142857}$ shown above.

Fractions allow us to carry out precise mathematical calculations easily with pen and paper, without the need for a calculator.

Example Calculate the sum of $\frac{1}{7}$ and $\frac{1}{3}$.

Let's say we decide, for reasons unknown, that it's a great day for decimal notation—we'd have to write our calculation as

$$\begin{aligned}
\text{ans} &= 0.\overline{142857} + 0.\overline{3} \\
&= 0.142\,857\,142\,857\ldots + 0.333\,333\,333\,333\ldots \\
&= 0.476\,190\,476\,190\,476\ldots \\
&= 0.\overline{476190}.
\end{aligned}$$

Wow that was complicated! This calculation is much simpler if we use fractions:

$$\frac{1}{7}+\frac{1}{3} = \frac{1}{7}\times\frac{3}{3} + \frac{1}{3}\times\frac{7}{7} = \frac{3}{21}+\frac{7}{21} = \frac{3+7}{21} = \frac{10}{21}.$$

Want to know how we did that? We multiplied the first term by $\frac{3}{3} = 1$ and the second term by $\frac{7}{7} = 1$ in order to obtain two equivalent fractions with the same denominator. This is one of the standard strategies when performing fraction operations: rewriting them as equivalent fractions that have the same denominator.

Equivalent fractions

The fractions $\frac{3}{8}$, $\frac{6}{16}$, and $\frac{12}{32}$ all correspond to the same number. Think about it—if you cut a pizza in 8 pieces and take 3 of them (see Figure 1.9), or you cut a pizza in 16 equal pieces and take 6 of them, you'll get the same amount of pizza in the end. All fractions of the form $\frac{3k}{8k}$ are *equivalent* to the fraction $\frac{3}{8}$, meaning they correspond to the same number.

Multiplying fractions

Fraction multiplication involves multiplying the numerators together and multiplying the denominators together:

$$\frac{a}{b} \times \frac{c}{d} = \frac{a \times c}{b \times d} = \frac{ac}{bd}.$$

For example, the product of the fractions $\frac{1}{3}$ and $\frac{1}{2}$ gives $\frac{1}{3} \times \frac{1}{2} = \frac{1 \times 1}{3 \times 2} = \frac{1}{6}$. This calculation shows that taking "one third of one half of some thing" is the same as taking one sixth of that thing.

Dividing fractions

To divide two fractions, compute the product of the first fraction times the second fraction "flipped" upside down:

$$\frac{a/b}{c/d} = \frac{a}{b} \div \frac{c}{d} = \frac{a}{b} \times \frac{d}{c} = \frac{a \times d}{b \times c} = \frac{ad}{bc}.$$

In other words, division by the fraction $\frac{c}{d}$ is the same as multiplication by the fraction $\frac{d}{c}$. This is called the "flip and multiply" rule for dividing fractions. For example, to divide the fraction $\frac{1}{3}$ by the fraction $\frac{1}{2}$, we do the following calculation: $\frac{1}{3}/\frac{1}{2} = \frac{1}{3} \times \frac{2}{1} = \frac{2}{3}$.

Reciprocals

The mathematical term *reciprocal* is used to describe the notion of "flipping" a number. The reciprocal of y is $\frac{1}{y}$, which is read "one over y." Multiplication by the reciprocal $\frac{1}{y}$ is the same as division by

y. The product of any number and its reciprocal equals one: $y \times \frac{1}{y} = \frac{y}{y} = 1$. The reciprocal of the fraction $\frac{m}{n}$ is the "flipped" fraction $\frac{n}{m}$. The product of $\frac{m}{n}$ and its reciprocal equals one: $\frac{m}{n} \times \frac{n}{m} = \frac{mn}{nm} = 1$.

Another way to denote the notion of "flipping" a number is to use the exponent negative one. The reciprocal of the number y is denoted y^{-1} and equals $\frac{1}{y}$. The reciprocal of the fraction $\frac{m}{n}$ is denoted $(\frac{m}{n})^{-1}$ and equals $\frac{n}{m}$. Using the negative exponent notation for reciprocals, we can write the "flip and multiply" rule for dividing fractions as $\frac{a}{b} \div \frac{c}{d} = \frac{a}{b} \times \left(\frac{c}{d}\right)^{-1} = \frac{a}{b} \times \frac{d}{c} = \frac{ad}{bc}$. We'll discuss negative exponents more generally in Section 3.1.

Adding fractions

Suppose we are asked to find the sum of the two fractions $\frac{a}{b}$ and $\frac{c}{d}$. If the denominators are the same, then we can add just the top parts: $\frac{1}{5} + \frac{2}{5} = \frac{3}{5}$. It makes sense to add the numerators since they refer to parts of the *same* whole.

However, if the denominators are different, we cannot add the numerators directly since they refer to parts of different wholes. Before we can add the numerators, we must rewrite the fractions so they have the same denominator, called a *common denominator*. We can obtain a common denominator by multiplying the first fraction by $\frac{d}{d} = 1$ and the second fraction by $\frac{b}{b} = 1$ in order to make the denominator of both fractions the same:

$$\frac{a}{b} + \frac{c}{d} = \frac{a}{b}\left(\frac{d}{d}\right) + \frac{c}{d}\left(\frac{b}{b}\right) = \frac{ad}{bd} + \frac{bc}{bd}.$$

Now that we have fractions with the same denominator, we can add their numerators. Note it's okay to change the denominator of a fraction as long as we also change the numerator in the same way. Multiplying the top and the bottom of a fraction by the same number is the same as multiplying by 1. So while the numbers of the fractions change, their equivalency is preserved:

$$\frac{a}{b} + \frac{c}{d} = \frac{ad}{bd} + \frac{bc}{bd} = \frac{ad + bc}{bd}.$$

Example To add $\frac{1}{6}$ and $\frac{1}{15}$, we can use the product of the two denominators as the common denominator, $6 \times 15 = 90$, and perform the fraction addition as follows:

$$\frac{1}{6} + \frac{1}{15} = \frac{1}{6} \times \frac{15}{15} + \frac{1}{15} \times \frac{6}{6} = \frac{15}{90} + \frac{6}{90} = \frac{21}{90} = \frac{7 \times \cancel{3}}{30 \times \cancel{3}} = \frac{7}{30}.$$

Note how we simplified the fraction by removing the factor 3 that is common to both the numerator and the denominator. The final answer $\frac{7}{30}$ is a reduced fraction equal to $\frac{21}{90}$.

The rules for subtracting fractions are the same as for adding fractions: computing the subtraction $\frac{a}{b} - \frac{c}{d}$ is the same as computing the addition $\frac{a}{b} + \frac{-c}{d}$.

Whole-and-fraction notation

A fraction greater than 1 like $\frac{5}{3}$ can also be denoted $1\frac{2}{3}$, which is read as "one and two-thirds." Similarly, $\frac{22}{7} = 3\frac{1}{7}$. We write the integer part of the number first, followed by the fractional part.

There is nothing wrong with writing fractions like $\frac{5}{3}$ and $\frac{22}{7}$. However, some teachers call these fractions *improper* and demand that all fractions are written in the whole-and-fraction way, as in $1\frac{2}{3}$ and $3\frac{1}{7}$. At the end of the day, both notations are correct.

Number line

The *number line* is a very useful visual representation for numbers. Every number from the sets $\mathbb{N}$, $\mathbb{Z}$, $\mathbb{Q}$, and $\mathbb{R}$ corresponds to some point on the number line. Developing a visual representation for numbers allows us to instantly compare the numbers' sizes based on their positions on the number line.

Figure 1.10 shows the natural numbers $\mathbb{N} = \{0, 1, 2, 3, 4, 5, \ldots\}$ represented as equally spaced points on the number line. We can construct the entire set of natural numbers by starting from 0 and taking steps of length one to the right on the number line. That's what counting is—we just keep adding one.

Note that natural numbers never end. We can always keep adding one to every number and obtain a larger number. The number line therefore extends to the right to infinity.

Figure 1.10: The natural numbers $\mathbb{N}$.

The integers $\mathbb{Z} = \{\ldots, -5, -4, -3, -2, -1, 0, 1, 2, 3, 4, 5, \ldots\}$ are similar to the naturals, but they also extend to the left of zero. Numbers to the left of zero are negative, while numbers to the right of zero are positive. The number line extends indefinitely on both sides, going to negative infinity on the left and positive infinity on the right.

Figure 1.11: The integers $\mathbb{Z}$.

The set of integers corresponds to a set of equally-spaced points on the number line with gaps of empty space between each integer. We need the real numbers $\mathbb{R}$ to fill these gaps.

The set of real numbers $\mathbb{R}$ is the complete representation of all possible points on the number line: every real number corresponds to some point on the number line, and every point on the number line corresponds to some real number. Visually, the set of real numbers fills the entire number line in **bold**, as shown in Figure 1.12. In other words, there are real numbers everywhere!

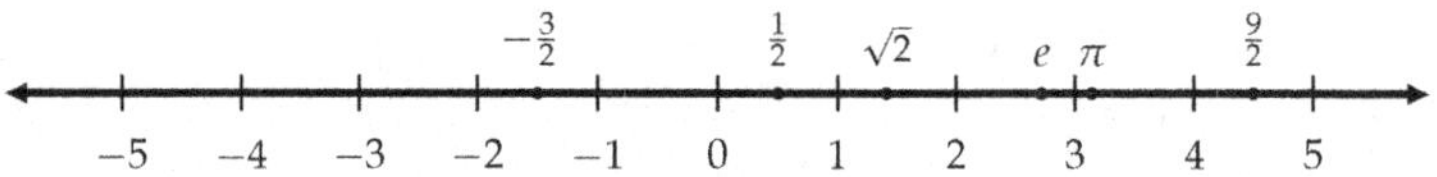

Figure 1.12: The real numbers $\mathbb{R}$ cover the entire number line.

Recall that the set of real numbers includes all the rational numbers like $-\frac{3}{2}$, $\frac{1}{2}$, and $\frac{9}{2}$, as well as irrational numbers like $\sqrt{2}$, e, and π. This means any number you are likely to run into when solving math problems can be visualized as a point on the number line. The number line can also be used to represent subsets of the real numbers. We'll talk about that in Section 8.3. For example, the subset of real numbers that are greater than two and smaller than four is shown in Figure 8.5 (page 175).

Discussion

Since we're still on the topic of number representations, I want to add some footnotes with "bonus material" related to the ideas we've covered in this section. Feel free to skip to the next section if you're in a hurry, because this is definitely not going to be on the exam!

Elementary arithmetic procedures

The four basic arithmetic operations are addition, subtraction, multiplication, and division. We can perform these operations for numerals a and b of any size using only pen and paper. It is sufficient to follow one of the well-defined procedures (called algorithms) for manipulating the individual digits that make up the numbers. The Wikipedia articles on elementary arithmetic and long division offer an excellent discussion of these procedures.

[Algorithms for performing elementary arithmetic]
`https://en.wikipedia.org/wiki/Elementary_arithmetic`
`https://en.wikipedia.org/wiki/Long_division`

Computer representations

Whenever you want to store a number on a computer, you must choose an appropriate computer representation for this number. The two most commonly used types of numbers in the computer world are integers (`int`) and floating point numbers (`float`). Computer integers can accurately describe the set of mathematical integers $\mathbb{Z}$, but there are limitations on the maximum size of numbers that computers can store. We can use floating point numbers to store decimals with up to 15 digits of precision. The `int` and `float` numbers that computers provide are sufficient for most practical computations, and you probably shouldn't worry about the limited precision of computer number representations. Still, I want you to be aware of the distinction between the abstract mathematical concept of a number and its computer representation. The real number $\sqrt{2}$ is irrational and has an infinite number of digits in its decimal expansion. On a computer, $\sqrt{2}$ is represented as the approximation 1.41421356237310 (a `float`). For most purposes the approximation is okay, but sometimes the limited precision can show up in calculations. For example, `float(sqrt(2))*float(sqrt(2)) = 2.0000000000000004` $\neq 2$ and `float(0.1)+float(0.2) = 0.30000000000000004` $\neq 0.3$. The result of the computer's calculation is only accurate up to the 15th digit. That's pretty good if you ask me.

Scientific notation

In science we often work with very large numbers like *the speed of light* (299 792 458), and very small numbers like *the permeability of free space* (0.000001256637). It can be difficult to judge the magnitude of such numbers and to carry out calculations on them using the usual decimal notation.

Dealing with such numbers is much easier if we use *scientific notation*. The speed of light can be written as 2.99792458×10^8, and the permeability of free space is denoted as 1.256637×10^{-6}. In both cases, we express the number as a decimal number between 1.0 and 9.9999... followed by the number 10 raised to some exponent. The effect of multiplying by 10^8 is to move the decimal point eight steps to the right, making the number bigger. Multiplying by 10^{-6} has the opposite effect, moving the decimal point to the left by six steps and making the number smaller. Scientific notation is useful because it allows us to clearly see the *size* of numbers: 1.23×10^6 is 1 230 000 whereas 1.23×10^{-10} is 0.000 000 000 123. With scientific notation you don't need to count the zeros!

The number of decimal places we use when specifying a certain physical quantity is usually an indicator of the *precision* with which we are able to measure this quantity. Taking into account the precision of the measurements we make is an important aspect of all quantitative research. Since elaborating further would be a digression, we won't go into a full discussion about the topic of *significant figures* here. Feel free to read the Wikipedia article on the subject to learn more.

Computer systems represent numbers using scientific notation, too. When entering a floating point number into the computer, separate the decimal part from the exponent by the character `e`, which stands for "exponent." For example, the speed of light is written as `2.99792458e8` and the permeability of free space is `1.256637e-6`.

Exercises

E1.4 Compute the value of the following expressions:

a) $\frac{1}{2} + \frac{1}{3}$ **b)** $\frac{1}{2} + \frac{1}{3} + \frac{1}{4}$ **c)** $3\frac{1}{2} + 2 - \frac{1}{3}$

Links

I encourage you to check the links provided below to learn more about numbers and number representations.

[History of the Hindu–Arabic system for representing numbers]
`https://en.wikipedia.org/wiki/Hindu-Arabic_numeral_system`

[Positional number representation systems]
`https://en.wikipedia.org/wiki/Positional_notation`

[Decimal representation]
`https://en.wikipedia.org/wiki/Decimal_representation`

[More details on scientific notation]
`https://en.wikipedia.org/wiki/Scientific_notation`

[Info about significant figures calculations]
`https://en.wikipedia.org/wiki/Significant_figures`

1.4 Variables

In math we use a lot of *variables* and *constants*, which are placeholder names for *any* number or unknown. Variables allow us to perform calculations without knowing all the details.

Example You're having tacos for lunch today and wondering how many you can eat without going over your caloric budget. Your goal is to eat 800 calories for lunch and you want to do the calculation before getting to the restaurant because you fear your math abilities might be affected in the presence of tacos. You're not sure how many calories each taco contains, so you invent the variable c to denote this unknown. You also define the variable x to represent the number of tacos you will eat, and come up with the equation $800 = cx$ to represent the total number of calories of your lunch. Solving for x, you find the total number of tacos you should order is $x = \frac{800}{c}$. If the restaurant serves tacos that contain $c = 200$ calories each, then you should order $x = \frac{800}{200} = 4$ of them. If the restaurant serves only giant tacos worth $c = 400$ calories each, then you can only eat $x = \frac{800}{400} = 2$ of them. Observe we were able to solve for x even before knowing the value of c.

Variable names

There are common naming patterns for variables:

- x: name used for the unknown in equations. We also use x to denote function inputs and the position of objects in physics.
- i, j, k, m, n: common names for integer variables
- a, b, c, d: letters near the beginning of the alphabet are often used to denote constants (fixed quantities that do not change).
- θ, ϕ: the Greek letters *theta* and *phi* are used to denote angles
- C: costs in business, along with P for profit, and R for revenue
- X: capital letters are used to denote random variables in probability theory

Variable substitution

We can often *change variables* and replace one unknown variable with another to simplify an equation. For example, say you don't feel comfortable around square roots. Every time you see a square root, you freak out until one day you find yourself taking an exam trying to solve for x in the following equation:

$$\frac{6}{5 - \sqrt{x}} = \sqrt{x}.$$

Don't freak out! In crucial moments like this, substitution can help with your root phobia. Just write, "Let $u = \sqrt{x}$" on your exam, and voila, you can rewrite the equation in terms of the variable u:

$$\frac{6}{5-u} = u,$$

which contains no square roots.

The next step to solve for u is to undo the division operation. Multiply both sides of the equation by $(5-u)$ to obtain

$$\frac{6}{5-u}(5-u) = u(5-u),$$

which simplifies to

$$6 = 5u - u^2.$$

This can be rewritten as the equation $u^2 - 5u + 6 = 0$, which in turn can be rewritten as $(u-2)(u-3) = 0$ using the techniques we'll learn in Section 2.2.

We now see that the solutions are $u_1 = 2$ and $u_2 = 3$. The last step is to convert our u-answers into x-answers by using $u = \sqrt{x}$, which is equivalent to $x = u^2$. The final answers are $x_1 = 2^2 = 4$ and $x_2 = 3^2 = 9$. Try plugging these x values into the original square root equation to verify that they satisfy it.

Compact notation

Symbolic manipulation is a powerful tool because it allows us to manage complexity. Say you're solving a physics problem in which you're told the mass of an object is $m = 140$ kg. If there are many steps in the calculation, would you rather use the number 140 kg in each step, or the shorter symbol m? It's much easier to use m throughout your calculation, and wait until the last step to substitute the value 140 kg when computing the final numerical answer.

1.5 Functions and their inverses

As we saw in the section on solving equations, the ability to "undo" functions is a key skill for solving equations.

Example Suppose we're solving for x in the equation

$$f(x) = c,$$

where f is some function and c is some constant. We're looking for the unknown x such that $f(x)$ equals c. Our goal is to isolate x on one side of the equation, but the function f stands in our way.

By using the *inverse function* (denoted f^{-1}) we "undo" the effects of f. We apply the inverse function f^{-1} to both sides of the equation to obtain

$$f^{-1}(f(x)) = f^{-1}(c).$$

By definition, the inverse function f^{-1} performs the opposite action of the function f, so together the two functions cancel each other out. We have $f^{-1}(f(x)) = x$ for any number x.

Provided everything is kosher (the function f^{-1} must be defined for the input c), the manipulation we made above is valid and we have obtained the answer $x = f^{-1}(c)$.

The above example introduces the notation f^{-1} for denoting the inverse function. This notation is inspired by the notation for reciprocals. Recall that multiplication by the reciprocal number a^{-1} is the inverse operation of multiplication by the number a: $a^{-1}ax = 1x = x$. In the case of functions, however, the negative-one exponent does not refer to "one over-$f(x)$" as in $\frac{1}{f(x)} = (f(x))^{-1}$; rather, it refers to the inverse function. In other words, the number $f^{-1}(y)$ is equal to the number x such that $f(x) = y$.

Be careful: sometimes an equation can have multiple solutions. For example, the function $f(x) = x^2$ maps two input values (x and $-x$) to the same output value $x^2 = f(x) = f(-x)$. The inverse function of $f(x) = x^2$ is $f^{-1}(y) = \sqrt{y}$, but both $x = +\sqrt{c}$ and $x = -\sqrt{c}$ are solutions to the equation $x^2 = c$. In this case, this equation's solutions can be indicated in shorthand notation as $x = \pm\sqrt{c}$.

Formulas

Here is a list of common functions and their inverses:

$$\begin{aligned}
\text{function } f(x) &\Leftrightarrow \text{inverse } f^{-1}(x) \\
x+2 &\Leftrightarrow x-2 \\
2x &\Leftrightarrow \tfrac{1}{2}x \\
-1x &\Leftrightarrow -1x \\
x^2 &\Leftrightarrow \pm\sqrt{x} \\
2^x &\Leftrightarrow \log_2(x) \\
3x+5 &\Leftrightarrow \tfrac{1}{3}(x-5) \\
a^x &\Leftrightarrow \log_a(x) \\
\exp(x) = e^x &\Leftrightarrow \ln(x) = \log_e(x) \\
\sin(x) &\Leftrightarrow \sin^{-1}(x) = \arcsin(x)
\end{aligned}$$

$$\cos(x) \quad \Leftrightarrow \quad \cos^{-1}(x) = \arccos(x)$$

The function-inverse relationship is *symmetric*—if you see a function on one side of the above table (pick a side, any side), you'll find its inverse on the opposite side.

Don't be surprised to see $-1x \Leftrightarrow -1x$ in the list of function inverses. Indeed, the opposite operation of multiplying by -1 is to multiply by -1 once more: $(-(-x) = x)$.

Example 1

If you want to solve the equation $x - 4 = 5$, you can apply the inverse function of $x - 4$, which is $x + 4$. After adding four to both sides of the equation, $x - 4 + 4 = 5 + 4$, we obtain the answer $x = 9$.

Example 2

Let's say your teacher doesn't like you and right away, on the first day of class, he gives you a serious equation and tells you to find x:

$$\log_5\left(3 + \sqrt{6\sqrt{x} - 7}\right) = 34 + \sin(8) - \Psi(1).$$

See what I mean when I say the teacher doesn't like you?

First, note that it doesn't matter what Ψ (the Greek letter *psi*) is, since x is on the other side of the equation. You can keep copying $\Psi(1)$ from line to line, until the end, when you throw the ball back to the teacher. "My answer is in terms of *your* variables, dude. *You* go figure out what the hell Ψ is since you brought it up in the first place!" By the way, it's not actually recommended to quote me verbatim should a situation like this arise. The same goes with sin(8). If you don't have a calculator handy, don't worry about it. Keep the expression sin(8) instead of trying to find its numerical value. In general, try to work with variables as much as possible and leave the numerical computations for the last step.

Okay, enough beating about the bush. Let's just find x and get it over with! On the right-hand side of the equation, we have the sum of a bunch of terms with no x in them, so we'll leave them as they are. On the left-hand side, the outermost function is a logarithm base 5. Cool. Looking at the table of inverse functions, we find the exponential function is the inverse of the logarithm: $a^x \Leftrightarrow \log_a(x)$. To get rid of $\log_5$, we must apply the exponential function base 5 to both sides:

$$5^{\log_5\left(3+\sqrt{6\sqrt{x}-7}\right)} = 5^{34+\sin(8)-\Psi(1)},$$

which simplifies to

$$3 + \sqrt{6\sqrt{x} - 7} = 5^{34+\sin(8)-\Psi(1)},$$

since 5^x cancels $\log_5 x$.

From here on, it is going to be as if Bruce Lee walked into a place with lots of bad guys. Addition of 3 is undone by subtracting 3 on both sides:

$$\sqrt{6\sqrt{x} - 7} = 5^{34+\sin(8)-\Psi(1)} - 3.$$

To undo a square root we take the square:

$$6\sqrt{x} - 7 = \left(5^{34+\sin(8)-\Psi(1)} - 3\right)^2.$$

Add 7 to both sides,

$$6\sqrt{x} = \left(5^{34+\sin(8)-\Psi(1)} - 3\right)^2 + 7,$$

divide by 6

$$\sqrt{x} = \frac{1}{6}\left(\left(5^{34+\sin(8)-\Psi(1)} - 3\right)^2 + 7\right),$$

and square again to find the final answer:

$$x = \left[\frac{1}{6}\left(\left(5^{34+\sin(8)-\Psi(1)} - 3\right)^2 + 7\right)\right]^2.$$

Did you see what I was doing in each step? Next time a function stands in your way, hit it with its inverse so it knows not to challenge you ever again.

Discussion

The recipe I have outlined above is not universally applicable. Sometimes x isn't alone on one side. Sometimes x appears in several places in the same equation. In these cases, you can't effortlessly work your way, Bruce Lee-style, clearing bad guys and digging toward x—you need other techniques.

The bad news is there's no general formula for solving complicated equations. The good news is the above technique of "digging toward the x" is sufficient for 80% of what you are going to be doing. You can get another 15% if you learn how to solve the quadratic equation:

$$ax^2 + bx + c = 0.$$

We'll show a formula for solving quadratic equations in Section 2.2. Solving cubic equations like $ax^3 + bx^2 + cx + d = 0$ using a formula is also possible, but at this point you might as well start using a computer to solve for the unknowns. See page 221 in Appendix C.

There are all kinds of other equations you can learn how to solve: equations with multiple variables, equations with logarithms, equations with exponentials, and equations with trigonometric functions. The principle of "digging" toward the unknown by applying inverse functions is the key for solving all these types of equations, so be sure to practice using it.

Exercises

E1.5 Solve for x in the following equations:

a) $3x = 6$ **b)** $\log_5(x) = 2$ **c)** $\log_{10}(\sqrt{x}) = 1$

E1.6 Find the function inverse and use it to solve the problems.

a) Solve the equation $f(x) = 4$, where $f(x) = \sqrt{x}$.

b) Solve for x in the equation $g(x) = 1$, given $g(x) = e^{-2x}$.

Chapter 2

Algebra

The rules of algebra tell us how to manipulate math expressions. By observing and working with the structure of expressions, we can simplify them, and thus compute their values more easily. In this chapter we'll learn the **general rules that apply to all math expressions**. Knowing these rules will equip you with a powerful toolbox of tricks you can apply to any problem you might face.

In Section 1.1, we learned how to *isolate* the unknown x in equations, which is by far the most important technique for solving equations. In this chapter we'll learn a few more techniques, including how to *simplify*, *expand*, and *factor* expressions. Together, these techniques are called *algebra*, from the Arabic *al-jabr* meaning "reunion of broken parts."

In Section 2.2, we'll use algebra to derive a general formula for solving quadratic equations of the form $ax^2 + bx + c = 0$. Knowing how to solve quadratic equations is really important because they show up all over the place.

Alright! Are you ready for algebra? Let's get started!

2.1 Basic rules of algebra

It's important that you know the general rules for manipulating numbers and variables, a process otherwise known as—you guessed it—*algebra*. This little refresher will cover these concepts to make sure you're comfortable on the algebra front. We'll also review some important algebraic tricks, like *factoring* and *completing the square*, which are useful when solving equations.

Let's define some terminology for referring to different parts of math expressions. When an expression contains multiple things added together, we call those things *terms*. Furthermore, terms are usually composed of many things multiplied together. When a number x is obtained as the product of other numbers like $x = abc$, we say "x factors into a, b, and c." We call a, b, and c the *factors* of x.

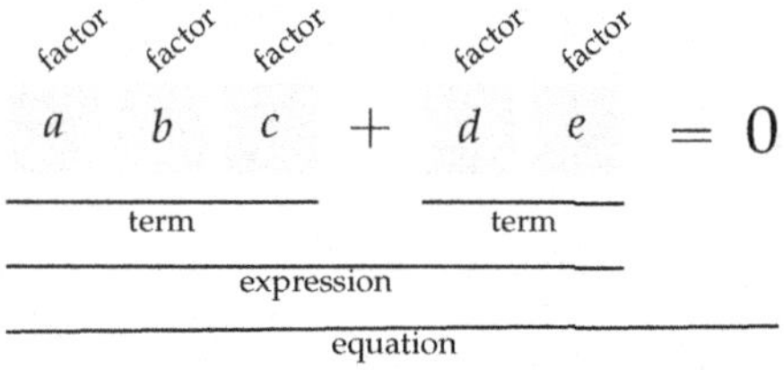

Figure 2.1: Diagram showing the names used to describe the different parts of the equation $abc + de = 0$.

Given any three numbers a, b, and c, we can apply the following algebraic properties:

1. Associative property: $a + b + c = (a + b) + c = a + (b + c)$ and $abc = (ab)c = a(bc)$
2. Commutative property: $a + b = b + a$ and $ab = ba$
3. Distributive property: $a(b + c) = ab + ac$

We use the distributive property every time we *expand* brackets. For example $a(b + c + d) = ab + ac + ad$. The brackets, also known as parentheses, indicate the expression $(b + c + d)$ must be treated as a whole; as a factor consisting of three terms. Multiplying this expression by a is the same as multiplying each term by a.

The opposite operation of expanding is called *factoring*, which consists of rewriting the expression with the common parts taken out in front of a bracket: $ab + ac = a(b + c)$. In this section, we'll discuss all algebra operations and illustrate what they're capable of.

Example Suppose we are asked to solve for t in the equation

$$7(3 + 4t) = 11(6t - 4).$$

Since the unknown t appears on both sides of the equation, it is not immediately obvious how to proceed.

To solve for t, we can bring all t terms to one side and all constant terms to the other side. First, expand the two brackets to obtain

$$21 + 28t = 66t - 44.$$

Then move things around to relocate all ts to the equation's right-hand side and all constants to the left-hand side:

$$21 + 44 = 66t - 28t.$$

We see t is contained in both terms on the right-hand side, so we can "factor it out" by rewriting the equation as

$$21 + 44 = t(66 - 28).$$

The answer is within close reach: $t = \frac{21+44}{66-28} = \frac{65}{38}$.

Expanding brackets

To *expand* a bracket is to multiply each term inside the bracket by the factor outside the bracket. The key thing to remember when expanding brackets is to apply the *distributive* property: $a(x + y) = ax + ay$. For longer expressions, we may need to apply the distributive property several times, until there are no more brackets left:

$$\begin{aligned}(a + b)(x + y + z) &= a(x + y + z) + b(x + y + z) \\ &= ax + ay + az + bx + by + bz.\end{aligned}$$

After expanding the brackets in this expression, we end up with six terms—one term for each of the six possible combinations of products between the terms in $(a + b)$ and the terms in $(x + y + z)$.

The distributive property is often used to manipulate expressions containing different powers of the variable x. For instance,

$$(x + 3)(x + 2) = x(x + 2) + 3(x + 2) = x^2 + x2 + 3x + 6.$$

We can use the commutative property on the second term $x2 = 2x$, then combine the two x terms into a single term to obtain

$$(x + 3)(x + 2) = x^2 + 5x + 6.$$

The bracket-expanding and simplification techniques demonstrated above are very common in math, and I recommend you solve some algebra practice problems to get the hang of them. Most math textbooks skip simplification steps and jump straight to the answer, since they assume readers are capable of doing simplifications on their own. It would be too long (and annoying) to show the simplifications for each expression. For example, the sentence "We can rewrite $(x + 3)(x + 2)$ as $x^2 + 5x + 6$," is the short version of the longer sentence, "We can apply the distributive property twice on $(x + 3)(x + 2)$ then combine the terms with the same power of x to get $x^2 + 5x + 6$."

It's not unusual for people to make math mistakes when expanding brackets and manipulating long algebra expressions. To avoid mistakes, use a step-by-step approach and apply only one operation in each step. Write legibly and keep the equations "organized" so it's easy to check the calculations performed in each step. Consider this slightly-more-complicated algebraic expression and its expansion:

$$\begin{aligned}(x+a)(bx^2+cx+d) &= x(bx^2+cx+d)+a(bx^2+cx+d)\\ &= bx^3+cx^2+dx \;+\; abx^2+acx+ad\\ &= bx^3+(c+ab)x^2+(d+ac)x+ad.\end{aligned}$$

Note how we sorted the terms in the final expression according to the different powers of x, with the terms containing x^2 grouped together, and the terms containing x grouped together. This approach helps keep things organized when dealing with expressions containing many terms.

Factoring

Factoring involves "taking out" the common parts of a complicated expression in order to make the expression more compact. Suppose we're given the expression $6x^2y+15x$. We can simplify this expression by taking out the common factors and moving them in front of a bracket. Let's see how to do this, step by step.

The expression $6x^2y+15x$ has two terms. Let's split each term into its constituent factors:

$$6x^2y+15x = (3)(2)(x)(x)y+(5)(3)x.$$

Since factors x and 3 appear in both terms, we can *factor them out* like this:

$$6x^2y+15x = 3x(2xy+5).$$

The expression on the right shows $3x$ is common to both terms.

Here's another example of factoring—notice the common factors are taken out and moved in front of the bracket:

$$2x^2y+2x+4x = 2x(xy+1+2) = 2x(xy+3).$$

Factoring quadratic expressions

A *quadratic expression* is an expression of the form ax^2+bx+c. The expression contains a *quadratic term* ax^2, a *linear term* bx, and a constant term c. The numbers a, b, and c are called *coefficients*: the quadratic coefficient is a, the linear coefficient is b, and the constant coefficient is c.

To *factor* the quadratic expression ax^2+bx+c is to rewrite it as the product of a constant and two factors like $(x+p)$ and $(x+q)$:

$$ax^2+bx+c = a(x+p)(x+q).$$

Rewriting quadratic expressions in factored form helps us better understand and describe their properties.

Example Suppose we're asked to describe the properties of the function $f(x) = x^2 - 5x + 6$. Specifically, we're asked to find the function's *roots*, which are the values of x for which the function equals zero.

Factoring the expression $x^2 - 5x + 6$ helps us see its properties more clearly, and makes our job of finding the roots of $f(x)$ easier. The factored form of this quadratic expression is

$$f(x) = x^2 - 5x + 6 = (x-2)(x-3).$$

Now we can see at a glance that the values of x for which $f(x) = 0$ are $x = 2$ and $x = 3$. When $x = 2$, the factor $(x-2)$ is zero and hence $f(x) = 0$. Similarly, when $x = 3$, the factor $(x-3)$ is zero so $f(x) = 0$.

How did we know that the factors of $x^2 - 5x + 6$ are $(x-2)$ and $(x-3)$ in the above example? For simple quadratics like the one above, we can simply *guess* the values of p and q in the equation $x^2 - 5x + 6 = (x+p)(x+q)$. Before we start guessing, let's look at the expanded version of the product between $(x+p)$ and $(x+q)$:

$$(x+p)(x+q) = x^2 + (p+q)x + pq.$$

Note the linear term on the right-hand side contains the sum of the unknowns $(p+q)$, while the constant term contains their product pq. If we want the equation $x^2 - 5x + 6 = x^2 + (p+q)x + pq$ to hold, we must find two numbers p and q whose sum equals -5 and whose product equals 6. After a couple of attempts we find $p = -2$ and $q = -3$. This guessing approach is an effective strategy for many of the factoring problems we will likely be asked to solve, since math teachers often choose simple numbers like ± 1, ± 2, ± 3, or ± 4 for the constants p and q. For more complicated quadratic expressions, we'll need to use the quadratic formula, which we'll talk about in Section 2.2.

Common quadratic forms

Let's look at some common variations of quadratic expressions you might encounter when doing algebra calculations.

The quadratic expression $x^2 - p^2$ is called a *difference of squares*, and it can be obtained by multiplying the factors $(x+p)$ and $(x-p)$:

$$(x+p)(x-p) = x^2 \cancel{-xp} \cancel{+px} - p^2 = x^2 - p^2.$$

There's no linear term because the $-xp$ term cancels the $+px$ term. Any time you see an expression like $x^2 - p^2$, you can know it comes from a product of the form $(x+p)(x-p)$.

A *perfect square* is a quadratic expression that can be written as the product of repeated factors $(x+p)$:

$$x^2 + 2px + p^2 = (x+p)(x+p) = (x+p)^2.$$

Note $x^2 - 2qx + q^2 = (x-q)^2$ is also a perfect square.

Completing the square

In this section we'll learn about an ancient algebra technique called *completing the square*, which allows us to rewrite *any* quadratic expression of the form $x^2 + Bx + C$ as a perfect square plus some constant correction factor $(x+p)^2 + k$. This algebra technique was described in one of the first books on *al-jabr* (algebra), written by Al-Khwarizmi around the year 800 CE. The name "completing the square" comes from the ingenious geometric construction used by this procedure. Yes, we can use geometry to solve algebra problems!

We assume the starting point for the procedure is a quadratic expression whose quadratic coefficient is one, $1x^2 + Bx + C$, and use capital letters B and C to denote the linear and constant coefficients. The capital letters are to avoid any confusion with the quadratic expression $ax^2 + bx + c$, for which $a \neq 1$. Note we can always write $ax^2 + bx + c$ as $a(x^2 + \frac{b}{a}x + \frac{c}{a})$ and apply the procedure to the expression inside the brackets, identifying $\frac{b}{a}$ with B and $\frac{c}{a}$ with C.

First let's rewrite the quadratic expression $x^2 + Bx + C$ by splitting the linear term into two equal parts:

$$x^2 + \tfrac{B}{2}x + \tfrac{B}{2}x + C.$$

We can interpret the first three terms geometrically as follows: the x^2 term corresponds to a square with side length x, while the two $\frac{B}{2}x$ terms correspond to rectangles with sides $\frac{B}{2}$ and x. See the left side of Figure 2.2 for an illustration.

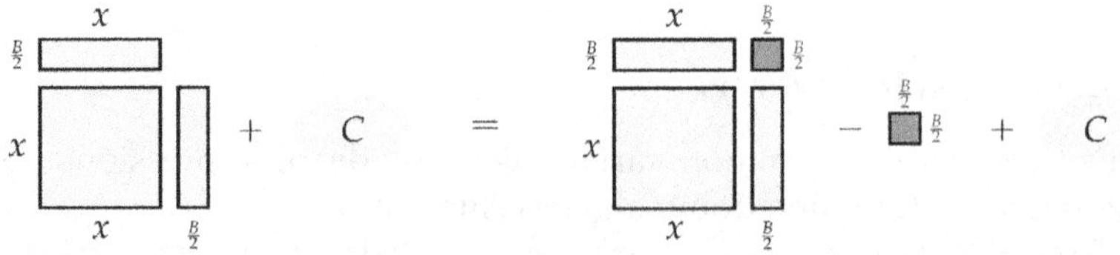

Figure 2.2: To complete the square in the expression $x^2 + Bx + C$, we need to add the quantity $(\frac{B}{2})^2$, which corresponds to a square (shown in darker colour) with sides equal to half the coefficient of the linear term. We also subtract $(\frac{B}{2})^2$ so the overall value of the expression remains unchanged.

The square with area x^2 and the two rectangles can be positioned to form a larger square with side length $\left(x+\frac{B}{2}\right)$. Note there's a small piece of sides $\frac{B}{2}$ by $\frac{B}{2}$ missing from the corner. To *complete the square*, we can add a term $\left(\frac{B}{2}\right)^2$ to this expression. To preserve the equality, we also subtract $\left(\frac{B}{2}\right)^2$ from the expression to obtain:

$$\begin{aligned} x^2+\tfrac{B}{2}x+\tfrac{B}{2}x+C &= \underbrace{x^2+\tfrac{B}{2}x+\tfrac{B}{2}x+\left(\tfrac{B}{2}\right)^2} - \left(\tfrac{B}{2}\right)^2 + C \\ &= \qquad \left(x+\tfrac{B}{2}\right)^2 \qquad - \left(\tfrac{B}{2}\right)^2 + C. \end{aligned}$$

The right-hand side of this equation describes the area of the square with side length $\left(x+\frac{B}{2}\right)$, minus the area of the small square $\left(\frac{B}{2}\right)^2$, plus the constant C, as illustrated on the right side of Figure 2.2.

We can summarize the entire procedure in one equation:

$$x^2+Bx+C = (x \underbrace{+\tfrac{B}{2}}_{(1)})^2 + C \underbrace{-\left(\tfrac{B}{2}\right)^2}_{(2)}.$$

There are two things to remember when you want to apply the complete-the-square trick: (1) choose the constant inside the bracket to be $\frac{B}{2}$ (half of the linear coefficient), and (2) subtract $\left(\frac{B}{2}\right)^2$ outside the bracket in order to keep the equation balanced.

Solving quadratic equations

Suppose we want to solve the quadratic equation $x^2+Bx+C=0$. It's not possible to solve this equation with the digging-toward-the-x approach from Section 1.1 (since x appears in both the quadratic term x^2 and the linear term Bx). Enter the completing-the-square trick!

Example Let's find the solutions of the equation $x^2+5x+6=0$. The coefficient of the linear term is $B=5$, so we choose $\frac{B}{2}=\frac{5}{2}$ for the constant inside the bracket, and subtract $\left(\frac{B}{2}\right)^2=\left(\frac{5}{2}\right)^2$ outside the bracket to keep the equation balanced. Completing the square gives

$$x^2+5x+6 = \left(x+\tfrac{5}{2}\right)^2+6-\left(\tfrac{5}{2}\right)^2 = 0.$$

Next we use fraction arithmetic to simplify the constant terms in the expression: $6-\left(\frac{5}{2}\right)^2 = 6\cdot\frac{4}{4}-\frac{25}{4}=\frac{24-25}{4}=\frac{-1}{4}=-0.25$.

We're left with the equation

$$(x+2.5)^2-0.25=0,$$

which we can now solve by digging toward x. First move 0.25 to the right-hand side to get $(x+2.5)^2 = 0.25$. Then take the square root on both sides to obtain $(x+2.5) = \pm 0.5$, which simplifies to $x = -2.5 \pm 0.5$. The two solutions are $x = -2.5 + 0.5 = -2$ and $x = -2.5 - 0.5 = -3$. You can verify these solutions by substituting the values in the original equation $(-2)^2 + 5(-2) + 6 = 0$ and similarly $(-3)^2 + 5(-3) + 6 = 0$. Congratulations, you just solved a quadratic equation using a 1200-year-old algebra technique!

In the next section, we'll learn how to leverage the complete-the-square trick to obtain a general-purpose formula for quickly solving quadratic equations.

Exercises

E2.1 Factor the following quadratic expressions:

a) $x^2 - 8x + 7$ **b)** $x^2 + 4x + 4$ **c)** $x^2 - 9$

Hint: Guess the values p and q in the expression $(x+p)(x+q)$.

E2.2 Solve the equations by completing the square.

a) $x^2 + 2x - 15 = 0$ **b)** $x^2 + 4x + 1 = 0$

2.2 Solving quadratic equations

What would you do if asked to solve for x in the quadratic equation $2x^2 = 4x + 6$? This is called a *quadratic equation* since it contains the unknown variable x squared. The name comes from the Latin *quadratus*, which means square. Quadratic equations appear often, so mathematicians created a general formula for solving them. In this section, we'll learn about this formula and use it to put some quadratic equations in their place.

Before we can apply the formula, we need to rewrite the equation we are trying to solve in the following form:

$$ax^2 + bx + c = 0.$$

This is called the *standard form* of the quadratic equation. We obtain this form by moving all the numbers and xs to one side and leaving only 0 on the other side. For example, to transform the quadratic equation $2x^2 = 4x + 6$ into standard form, we subtract $4x + 6$ from both sides of the equation to obtain $2x^2 - 4x - 6 = 0$. What are the values of x that satisfy this equation?

Quadratic formula

The solutions to the equation $ax^2 + bx + c = 0$ for $a \neq 0$ are

$$x_1 = \frac{-b + \sqrt{b^2 - 4ac}}{2a} \qquad \text{and} \qquad x_2 = \frac{-b - \sqrt{b^2 - 4ac}}{2a}.$$

The quadratic formula is usually abbreviated $x = \frac{-b \pm \sqrt{b^2 - 4ac}}{2a}$, where the sign "$\pm$" stands for both "$+$" and "$-$." The notation "$\pm$" allows us to express both solutions x_1 and x_2 in one equation, but you should keep in mind there are really two solutions.

Let's see how the quadratic formula is used to solve the equation $2x^2 - 4x - 6 = 0$. Finding the two solutions requires the simple mechanical task of identifying $a = 2$, $b = -4$, and $c = -6$, then plugging these values into the two parts of the formula:

$$x_1 = \frac{4 + \sqrt{4^2 - 4(2)(-6)}}{4} = \frac{4 + \sqrt{16 + 48}}{4} = \frac{4 + \sqrt{64}}{4} = 3,$$

$$x_2 = \frac{4 - \sqrt{4^2 - 4(2)(-6)}}{4} = \frac{4 - \sqrt{16 + 48}}{4} = \frac{4 - \sqrt{64}}{4} = -1.$$

We can easily verify that value $x_1 = 3$ and $x_2 = -1$ both satisfy the original equation $2x^2 = 4x + 6$.

Proof of the quadratic formula

Every claim made by a mathematician comes with a *proof*, which is a step-by-step argument that shows why the claim is true. It's easy to see where a proof starts and where a proof ends in mathematical texts. Each proof begins with the heading *Proof* (usually in italics) and has the symbol "□" at its end. The purpose of these demarcations is to give readers the option to skip the proof. It's not necessary to read and understand the proofs of all math statements, but reading proofs can often lead you to a more solid understanding of the material.

I want you to see the proof of the quadratic formula because it's an important result that you'll use very often to solve math problems. Reading the proof will help you understand where the quadratic formula comes from. The proof relies on the completing-the-square technique from the previous section, and some basic algebra operations. You can totally handle this!

Proof. We're starting from the quadratic equation $ax^2 + bx + c = 0$, and we're making the additional assumption that $a \neq 0$. We want to find the value or values of x that satisfy this equation.

The first thing we want to do is divide by a to obtain the equivalent equation

$$x^2 + \frac{b}{a}x + \frac{c}{a} = 0.$$

We are allowed to divide by a since we assumed that $a \neq 0$.

Next we apply the *complete the square* trick to the quadratic expression, to obtain an equivalent expression of the form $(x+?)^2+?$. Recall that the trick for completing the square is to choose the number inside the bracket to be half the coefficient of the linear term of the quadratic expression, which is $\frac{b}{2a}$ in this case. We must also subtract the square of this term outside the bracket in order to maintain the equality. After completing the square, we're left with the following equation:

$$\left(x + \frac{b}{2a}\right)^2 + \frac{c}{a} - \frac{b^2}{4a^2} = 0.$$

From here, we use the standard digging-toward-the-x procedure. Move all constants to the right-hand side,

$$\left(x + \frac{b}{2a}\right)^2 = \frac{b^2}{4a^2} - \frac{c}{a},$$

and take the square root of both sides to undo the square function:

$$x + \frac{b}{2a} = \pm\sqrt{\frac{b^2}{4a^2} - \frac{c}{a}}.$$

Since any number and its opposite have the same square, taking the square root gives us two possible solutions, which we denote using the "$\pm$" symbol.

Next we subtract $\frac{b}{2a}$ from both sides of the equation to isolate x and obtain $x = -\frac{b}{2a} \pm \sqrt{\frac{b^2}{4a^2} - \frac{c}{a}}$. We tidy up the mess under the square root, $\sqrt{\frac{b^2}{4a^2} - \frac{c}{a}} = \sqrt{\frac{b^2}{4a^2} - \frac{4a \cdot c}{4a \cdot a}} = \sqrt{\frac{b^2-4ac}{4a^2}} = \frac{\sqrt{b^2-4ac}}{2a}$, and add the fractions on the right-hand side to obtain $x = \frac{-b \pm \sqrt{b^2-4ac}}{2a}$. The solutions to the quadratic equation $ax^2 + bx + c = 0$ are

$$x_1 = \frac{-b + \sqrt{b^2 - 4ac}}{2a} \qquad \text{and} \qquad x_2 = \frac{-b - \sqrt{b^2 - 4ac}}{2a}.$$

This completes the proof of the quadratic formula. □

The expression $b^2 - 4ac$ is called the *discriminant* of the equation. The discriminant tells us important information about the solutions of the equation $ax^2 + bx + c = 0$. The solutions x_1 and x_2 correspond to real numbers if the discriminant is positive or zero: $b^2 - 4ac \geqslant 0$. When the discriminant is zero ($b^2 - 4ac = 0$), the equation has only one solution since $x_1 = x_2 = \frac{-b}{2a}$. If the discriminant is negative, $b^2 - 4ac < 0$, the quadratic formula requires computing the square root of a negative number, which is not allowed for real numbers.

Alternative proof

To prove the quadratic formula, we don't necessarily need to show the algebra steps we followed to obtain the formula as outlined above. The quadratic formula states that x_1 and x_2 are solutions. To prove the formula is correct we can simply plug x_1 and x_2 into the equation $ax^2 + bx + c = 0$ to verify that x_1 and x_2 are solutions. Verify this on your own.

Applications

The golden ratio

The *golden ratio* is an essential proportion in geometry, art, aesthetics, biology, and mysticism, and is usually denoted as $\varphi = \frac{1+\sqrt{5}}{2} = 1.6180339\ldots$. This ratio is determined as the positive solution to the quadratic equation

$$x^2 - x - 1 = 0.$$

Applying the quadratic formula to this equation yields two solutions,

$$x_1 = \frac{1+\sqrt{5}}{2} = \varphi \qquad \text{and} \qquad x_2 = \frac{1-\sqrt{5}}{2} = -\frac{1}{\varphi}.$$

You can learn more about the various contexts in which the golden ratio appears from the Wikipedia article on the subject.

Explanations

Multiple solutions

Often, we are interested in only one of the two solutions to the quadratic equation. It will usually be obvious from the context of the problem which of the two solutions should be kept and which should be discarded. For example, the *time of flight* of a ball thrown in the air from a height of 3 metres with an initial velocity of 12 metres per second is obtained by solving the equation $(-4.9)t^2 + 12t + 3 = 0$. The two solutions of the quadratic equation are $t_1 = -0.229$ and $t_2 = 2.678$. The first answer t_1 corresponds to a time in the past so we reject it as invalid. The correct answer is t_2. The ball will hit the ground after $t = 2.678$ seconds.

Relation to factoring

In the previous section we discussed the *quadratic factoring* operation by which we could rewrite a quadratic function as the product of a constant and two factors:

$$f(x) = ax^2 + bx + c = a(x - x_1)(x - x_2).$$

The two numbers x_1 and x_2 are called the *roots* of the function: these points are where the function $f(x)$ touches the x-axis.

You now have the ability to factor any quadratic equation: use the quadratic formula to find the two solutions, x_1 and x_2, then rewrite the expression as $a(x - x_1)(x - x_2)$.

Some quadratic expressions cannot be factored, however. These "unfactorable" expressions correspond to quadratic functions whose graphs do not touch the x-axis. They have no real solutions (no roots). There is a quick test you can use to check if a quadratic function $f(x) = ax^2 + bx + c$ has roots (touches or crosses the x-axis) or doesn't have roots (never touches the x-axis). If $b^2 - 4ac > 0$ then the function f has two roots. If $b^2 - 4ac = 0$, the function has only one root, indicating the special case when the function touches the x-axis at only one point. If $b^2 - 4ac < 0$, the function has no roots. In this case, the quadratic formula fails because it requires taking the square root of a negative number, which is not allowed (for now). We'll come back to the idea of taking square roots of negative numbers in Section 7.5 (see page 153).

Links

[Algebra explanation of the quadratic formula]
`https://www.youtube.com/watch?v=r3SEkdtpobo`

[Visual explanation of the quadratic formula derivation]
`https://www.youtube.com/watch?v=EBbtoFMJvFc`

Exercises

E2.3 Solve for x in the quadratic equation $2x^2 - x = 3$.

E2.4 Solve for x in the equation $x^4 - 4x^2 + 4 = 0$.
Hint: Use the substitution $y = x^2$.

Chapter 3

Exponents and logarithms

Understanding exponents and logarithms is important for your math modelling skills. Exponential functions can describe population growth, interest calculations on loans, probabilities of random events, voltages in electric capacitors, radioactive decay, and many other phenomena.

The logarithmic function is the inverse of the exponential function. We need the logarithmic function every time we want to solve an equation containing exponents. Additionally, the logarithmic scale allows us to conveniently compare numbers of vastly different sizes. Logarithmic scales are used to describe systems with exponential scale differences, including sound intensity and the acidity of chemical solutions.

As a student of math, you can consider exponentiation and logarithm computation as basic operations, on the same footing as addition, subtraction, multiplication, and division.

$$\text{Basic math operations: } \{\ x + y,\ x - y,\ xy,\ x/y,\ x^y,\ \log_x y\ \}.$$

In this chapter, you'll learn the basic rules for manipulating exponents and logarithms. As you read and practice, your job is to make sure you're just as comfortable with exponents as you are with addition and multiplication, and as comfortable with logarithms as you are with subtraction and division.

3.1 Exponents

In math we must often multiply together the same number many times, so we use the notation

$$b^n = \underbrace{bbb \cdots bb}_{n \text{ times}}$$

to denote some number b multiplied by itself n times. In this section we'll review the basic terminology associated with exponents and discuss their properties.

Definitions

The fundamental ideas of exponents are:

- b^n: the number b raised to the power n
  - ▷ b: the *base*
  - ▷ n: the *exponent* or *power* of b in the expression b^n

By definition, the zeroth power of any number is equal to one, expressed as $b^0 = 1$.

We'll also discuss *exponential functions*. In particular, we define the following important exponential functions:

- b^x: the exponential function base b
- 10^x: the exponential function base 10
- $\exp(x) = e^x$: the exponential function base e. The number e is called *Euler's number*.
- 2^x: the exponential function base 2. This function is important in computer science.

The number $e = 2.7182818\ldots$ is a special base with many applications. We call e the *natural* base. Another special base is 10 because we use the decimal system for our numbers. We can write very large numbers and very small numbers as powers of 10. For example, one thousand can be written as $1\,000 = 10^3$, one million is $1\,000\,000 = 10^6$, and one billion is $1\,000\,000\,000 = 10^9$.

Formulas

The following properties follow from the definition of exponentiation as repeated multiplication.

Property 1 Multiplying together two exponential expressions that have the same base is the same as adding the exponents:

$$b^m b^n = \underbrace{bbb\cdots bb}_{m \text{ times}}\underbrace{bbb\cdots bb}_{n \text{ times}} = \underbrace{bbbbbb\cdots bb}_{m+n \text{ times}} = b^{m+n}.$$

Property 2 Division by a number can be expressed as an exponent of minus one:

$$b^{-1} = \frac{1}{b}.$$

Any number times its reciprocal gives one: $bb^{-1} = \frac{b}{b} = 1$. A negative exponent corresponds to a division:

$$b^{-n} = \frac{1}{b^n}.$$

Property 3 By combining Property 1 and Property 2 we obtain the following rule:

$$\frac{b^m}{b^n} = b^{m-n}.$$

In particular we have $b^n b^{-n} = b^{n-n} = b^0 = 1$. Multiplication by the number b^{-n} is the inverse operation of multiplication by the number b^n. The net effect of the combination of both operations is the same as multiplying by one.

Property 4 When an exponential expression is exponentiated, the inner exponent and the outer exponent multiply:

$$(b^m)^n = \underbrace{(\underbrace{bbb\cdots bb}_{m \text{ times}})(\underbrace{bbb\cdots bb}_{m \text{ times}})\cdots(\underbrace{bbb\cdots bb}_{m \text{ times}})}_{n \text{ times}} = b^{mn}.$$

Property 5.1

$$(ab)^n = \underbrace{(ab)(ab)(ab)\cdots(ab)(ab)}_{n \text{ times}} = \underbrace{aaa\cdots aa}_{n \text{ times}}\underbrace{bbb\cdots bb}_{n \text{ times}} = a^n b^n.$$

Property 5.2

$$\left(\frac{a}{b}\right)^n = \underbrace{\left(\frac{a}{b}\right)\left(\frac{a}{b}\right)\left(\frac{a}{b}\right)\cdots\left(\frac{a}{b}\right)\left(\frac{a}{b}\right)}_{n \text{ times}} = \frac{\overbrace{aaa\cdots aa}^{n \text{ times}}}{\underbrace{bbb\cdots bb}_{n \text{ times}}} = \frac{a^n}{b^n}.$$

Property 6 Raising a number to the power $\frac{1}{n}$ is equivalent to finding the nth root of the number:

$$b^{\frac{1}{n}} = \sqrt[n]{b}.$$

In particular, the square root corresponds to the exponent of one half: $\sqrt{b} = b^{\frac{1}{2}}$. The cube root (the inverse of x^3) corresponds to $\sqrt[3]{b} = b^{\frac{1}{3}}$. We can verify the inverse relationship between $\sqrt[3]{x}$ and x^3 by using either Property 1: $(\sqrt[3]{x})^3 = (x^{\frac{1}{3}})(x^{\frac{1}{3}})(x^{\frac{1}{3}}) = x^{\frac{1}{3}+\frac{1}{3}+\frac{1}{3}} = x^1 = x$, or by using Property 4: $(\sqrt[3]{x})^3 = (x^{\frac{1}{3}})^3 = x^{\frac{3}{3}} = x^1 = x$.

Properties 5.1 and 5.2 also apply for fractional exponents:

$$\sqrt[n]{ab} = (ab)^{\frac{1}{n}} = a^{\frac{1}{n}} b^{\frac{1}{n}} = \sqrt[n]{a}\sqrt[n]{b}, \qquad \sqrt[n]{\left(\frac{a}{b}\right)} = \left(\frac{a}{b}\right)^{\frac{1}{n}} = \frac{a^{\frac{1}{n}}}{b^{\frac{1}{n}}} = \frac{\sqrt[n]{a}}{\sqrt[n]{b}}.$$

Discussion

Negative exponents

A negative sign in the exponent does not mean "subtract," but rather "divide by":

$$a^{-n} = \frac{1}{a^n} = \frac{1}{\underbrace{aaa\cdots a}_{n\text{ times}}}.$$

To understand why negative exponents correspond to division, consider the following calculation:

$$a^m a^n = \underbrace{aaa\cdots a}_{m\text{ times}}\ \underbrace{aaa\cdots a}_{n\text{ times}} = \underbrace{aaaaaa\cdots\cdots aa}_{m+n\text{ times}} = a^{m+n}.$$

This calculation illustrates a general rule for multiplying exponential expressions: $a^m a^n = a^{m+n}$, or, if you prefer words, "add the exponents together when multiplying exponential expressions." Defining $a^{-n} = \frac{1}{a^n}$ ensures the rule $a^m a^n = a^{m+n}$ is also valid for negative exponents:

$$a^m a^{-n} = \underbrace{aaaaa\cdots aa}_{m\text{ times}} \frac{1}{\underbrace{a\cdots a}_{n\text{ times}}} = \frac{\overbrace{aaaaa\cdots aa}^{m\text{ times}}}{\underbrace{a\cdots a}_{n\text{ times}}} = \underbrace{aa\cdots a}_{m-n\text{ times}} = a^{m-n}.$$

For example, the expression 2^{-3} corresponds to $\frac{1}{2^3} = \frac{1}{8}$. If we multiply together 2^5 and 2^{-3}, we obtain $2^5 \cdot 2^{-3} = 2^{5-3} = 2^2 = 4$.

Fractional exponents

We discussed positive and negative exponents, but what about exponents that are fractions? Fractional exponents describe square-root-like operations:

$$a^{\frac{1}{2}} = \sqrt{a} = \sqrt[2]{a}, \qquad a^{\frac{1}{3}} = \sqrt[3]{a}, \qquad a^{\frac{1}{4}} = \sqrt[4]{a}.$$

Recall the square-root operation $\sqrt{\ }$, which is used it to undo the effect of the x^2 operation. More generally, the "n^{th} root" function $\sqrt[n]{x}$ is the inverse of the function x^n.

If this is the first time you're seeing square roots, you might assume you'll need to learn lots of new rules for manipulating square-root expressions. Or maybe you have experience with the rules of "squiggle math" already. What kind of emotions do expressions like $\sqrt[3]{27}\sqrt[3]{8}$ stir up in you? Chill! There's no new math to learn and no rules to memorize. All the "squiggle math" rules are consequences of the general rule $a^b a^c = a^{b+c}$ applied to expressions where the exponents are fractions. For example, a cube root satisfies the equation

$$\sqrt[3]{a}\sqrt[3]{a}\sqrt[3]{a} = a^{\frac{1}{3}}a^{\frac{1}{3}}a^{\frac{1}{3}} = a^{\frac{1}{3}+\frac{1}{3}+\frac{1}{3}} = a^1 = a.$$

Do you see why $\sqrt[3]{x}$ and x^3 are inverse operations? The number $\sqrt[3]{a}$ is one third of the number a with respect to multiplication (since multiplying $\sqrt[3]{a}$ by itself three times produces a). We say "one third *with respect to multiplication*," because the usual meaning of "one third of a" is with respect to the addition operation (adding together three copies of $\frac{a}{3}$ produces a).

The nth root of a is a number which, when multiplied together n times, will give a. The "n^{th} root of a" can be denoted in two equivalent ways:

$$\sqrt[n]{a} = a^{\frac{1}{n}}.$$

Using this definition and the general rule $a^b a^c = a^{b+c}$ allows us to simplify all kinds expressions. For example, we can simplify $\sqrt[4]{a}\sqrt[4]{a}$ by rewriting it as $\sqrt[4]{a}\sqrt[4]{a} = a^{\frac{1}{4}}a^{\frac{1}{4}} = a^{\frac{1}{4}+\frac{1}{4}} = a^{\frac{1}{2}} = \sqrt{a}$. We can also simplify the expression $\sqrt[3]{27}\sqrt[3]{8}$ by rewriting it as $27^{\frac{1}{3}}8^{\frac{1}{3}}$, then simplifying it as $27^{\frac{1}{3}}8^{\frac{1}{3}} = (3\cdot3\cdot3)^{\frac{1}{3}}\,(2\cdot2\cdot2)^{\frac{1}{3}} = 3\cdot2 = 6$.

Let's verify the claim that $\sqrt[n]{a}$ equals "one n^{th} of a with respect to multiplication." To obtain the whole number, we must multiply the number $\sqrt[n]{a}$ times itself n times:

$$\left(\sqrt[n]{a}\right)^n = \left(a^{\frac{1}{n}}\right)^n = \underbrace{a^{\frac{1}{n}}a^{\frac{1}{n}}a^{\frac{1}{n}}a^{\frac{1}{n}}\cdots a^{\frac{1}{n}}a^{\frac{1}{n}}}_{n \text{ times}} = a^{\frac{n}{n}} = a^1 = a.$$

The n-fold product of $\frac{1}{n}$-fractional exponents of any number produces that number raised to exponent one, and therefore the inverse operation of $\sqrt[n]{x}$ is x^n.

The commutative law of multiplication $ab = ba$ implies that we can write any fraction $\frac{a}{b}$ in two other equivalent ways: $\frac{a}{b} = a\frac{1}{b} = \frac{1}{b}a$. We multiply by a, then divide the result by b; or first we divide by b and then multiply the result by a. Similarly, when we have a fraction in the exponent, we can write the answer in two equivalent ways:

$$a^{\frac{2}{3}} = \sqrt[3]{a^2} = (\sqrt[3]{a})^2, \qquad a^{-\frac{1}{2}} = \frac{1}{a^{\frac{1}{2}}} = \frac{1}{\sqrt{a}}, \qquad a^{\frac{m}{n}} = \left(\sqrt[n]{a}\right)^m = \sqrt[n]{a^m}.$$

Make sure the above notation makes sense to you. As an exercise, try computing $5^{\frac{4}{3}}$ on your calculator and check that you obtain $8.54987973\ldots$ as the answer.

Even and odd exponents

The function $f(x) = x^n$ behaves differently depending on whether the exponent n is even or odd. If n is odd we have

$$\left(\sqrt[n]{b}\right)^n = \sqrt[n]{b^n} = b, \qquad \text{when } n \text{ is odd.}$$

However, if n is even, the function x^n destroys the sign of the number (see x^2, which maps both $-x$ and x to x^2). The successive application of exponentiation by n and the nth root has the same effect as the absolute value function:

$$\sqrt[n]{b^n} = |b|, \qquad \text{when } n \text{ is even.}$$

Recall that the absolute value function $|x|$ discards the information about the sign of x. The expression $(\sqrt[n]{b})^n$ cannot be computed whenever b is a negative number. The reason is that we can't evaluate $\sqrt[n]{b}$ for $b < 0$ in terms of real numbers, since there is no real number which, multiplied by itself an even number of times, gives a negative number.

Links

[Further reading on exponentiation]
`http://en.wikipedia.org/wiki/Exponentiation`

Exercises

E3.1 Simplify the following exponential expressions.

a) $2^3 ef \frac{\sqrt{ef}}{(\sqrt{ef})^3}$ **b)** $\frac{abc}{a^2b^3c^4}$ **c)** $\frac{(2\alpha)^3}{\alpha}$ **d)** $(a^3)^2(\frac{1}{b})^2$

E3.2 Simplify the following expressions as much as possible:

a) $\sqrt{3}\sqrt{3}$ **b)** $\sqrt{9}\sqrt{16}$ **c)** $\frac{\sqrt[3]{8}\sqrt{3}}{\sqrt{4}}$ **d)** $\frac{\sqrt{aba}}{\sqrt{b}}$

E3.3 Calculate the values of the following exponential expressions:

a) $\sqrt{2}(\pi)2^{\frac{1}{2}}$ **b)** $8^{\frac{2}{3}} + 8^{-\frac{2}{3}}$ **c)** $\left(\frac{(\sqrt[3]{c})^3}{c}\right)^{77}$ **d)** $\left(\frac{x^2\sqrt{x^4}}{x^3}\right)^2$

E3.4 Find all the values of x that satisfy these equations:

a) $x^2 = a$ **b)** $x^3 = b$ **c)** $x^4 = c$ **d)** $x^5 = d$

E3.5 Coulomb's constant k_e is defined by the formula $k_e = \frac{1}{4\pi\varepsilon_0}$, where ε_0 is the permittivity of free space. Use a calculator to compute the value of k_e starting from $\varepsilon_0 = 8.854 \times 10^{-12}$ and $\pi = 3.14159265$. Report your answer with an appropriate number of digits, even if the calculator gives you a number with more digits.

3.2 Logarithms

Some people think the word "logarithm" refers to some mythical, mathematical beast. Legend has it that logarithms are many-headed, breathe fire, and are extremely difficult to understand. Nonsense! Logarithms are simple. It will take you at most a couple of pages to get used to manipulating them, and that is a good thing because logarithms are used all over the place.

The strength of your sound system is measured in logarithmic units called decibels. This is because your ears are sensitive only to exponential differences in sound intensity. Logarithms allow us to compare very large numbers and very small numbers on the same scale. If sound were measured in linear units instead of logarithmic units, your sound system's volume control would need to range from 1 to 1 048 576. That would be weird, no? This is why we use the logarithmic scale for volume notches. Using a logarithmic scale, we can go from sound intensity level 1 to sound intensity level 1 048 576 in 20 "progressive" steps. Assume each notch doubles the sound intensity, rather than increasing the intensity by a fixed amount. If the first notch corresponds to 2, the second notch is 4—still probably inaudible, turn it up! By the time you get to the sixth notch you're at $2^6 = 64$ sound intensity, which is the level of audible music. The tenth notch corresponds to sound intensity $2^{10} = 1024$ (medium-strength sound), and finally the twentieth notch reaches a max power of $2^{20} = 1\,048\,576$, at which point the neighbours come to complain.

Definitions

You're hopefully familiar with these following concepts from the previous section:

- b^x: the exponential function base b
- $\exp(x) = e^x$: the exponential function base e, Euler's number
- 2^x: exponential function base 2
- f: the notion of a function
- f^{-1}: the inverse function of f. The inverse function is defined in terms of f such that $f^{-1}(f(x)) = x$. In other words, if you apply f to some number x and get the output y, and then you pass y through f^{-1}, the output will be x again. The inverse function f^{-1} undoes the effects of the function f.

In this section we'll play with the following new concepts:

- $\log_b(x)$: the logarithm of x base b is the inverse function of b^x.
- $\ln(x)$: the "natural" logarithm base e. This is the inverse of e^x.
- $\log_2(x)$: the logarithm base 2 is the inverse of 2^x.

I say *play* because there is nothing much new to learn here: a logarithm is a clever way to talk about the size of a number; essentially, it tells us how many digits the number has.

Formulas

The main thing to realize is that logs don't really exist on their own. They are defined as the inverses of their corresponding exponential functions. The following statements are equivalent:

$$\log_b(x) = m \qquad \Leftrightarrow \qquad b^m = x.$$

Logarithms with base e are written $\ln(x)$ for "logarithme naturel" because e is the "natural" base. Another special base is 10 because our numbers are based on the decimal system. The logarithm base 10 $\log_{10}(x)$ tells us roughly the size of the number x—how many digits the number has.

Example When someone working for the System (say someone with a high-paying job in the financial sector) boasts about his or her "six-figure" salary, they are really talking about the log of how much money they make. The "number of figures" N_S in their salary is calculated as 1 plus the logarithm base 10 of their salary S. The formula is

$$N_S = 1 + \log_{10}(S).$$

A salary of $S = 100\,000$ corresponds to $N_S = 1 + \log_{10}(100\,000) = 1 + 5 = 6$ figures.

What is the smallest "seven-figure" salary? We must solve for S given $N_S = 7$ in the formula. We find $7 = 1 + \log_{10}(S)$, which means $6 = \log_{10}(S)$, and—using the inverse relationship between logarithm base 10 and exponentiation base 10—we discover $S = 10^6 = 1\,000\,000$. One million dollars per year! Yes, for this kind of money I see how someone might want to work for the System. But most system pawns never make it to the seven-figure level; I believe the average high-ranking salary is more in the $1 + \log_{10}(250\,000) = 1 + 5.397 = 6.397$ digits range. Wait, a lousy 0.397 extra digits is all it takes to convince some of the smartest people out there to sell their brains to the finance sector? What wankers! Who needs a six-digit salary anyway? Why not make $1 + \log_{10}(55\,000) = 5.74$ digits as a teacher and do something with your life that *actually* matters?

Properties

Moving on, let's discuss two important properties you'll need when dealing with logarithms. Pay attention because the arithmetic rules for logarithms are very different from the usual rules for numbers. Intuitively, you can think of logarithms as a convenient way to refer to the exponents of numbers. The following properties are the logarithmic analogues of the properties of exponents.

Property 1

The first property states that the sum of the logarithms of two numbers is equal to the logarithm of the product of the numbers:

$$\log(x) + \log(y) = \log(xy).$$

Proof. We need to show that the expression on the left is equal to the expression on the right. We met logarithms very recently, so we don't know each other too well yet. In fact, the only thing we know about logs is the inverse relationship with the exponential function. The only way to prove this property is to use this relationship.

The following statement is true for any base b:

$$b^m b^n = b^{m+n}.$$

This follows from first principles. Recall that exponentiation is nothing more than repeated multiplication. If you count the total number of bs multiplied on the left side, you'll find a total of $m + n$ of them, which is what we have on the right.

If we define some new variables x and y such that $b^m = x$ and $b^n = y$, then we can rewrite the equation $b^m b^n = b^{m+n}$ as

$$xy = b^{m+n}.$$

Taking the logarithm of both sides gives us

$$\log_b(xy) = \log_b\left(b^{m+n}\right) = m + n = \log_b(x) + \log_b(y).$$

The last step above uses the definition of the log function again, which states that

$$b^m = x \quad \Leftrightarrow \quad m = \log_b(x) \qquad \text{and} \qquad b^n = y \quad \Leftrightarrow \quad n = \log_b(y).$$

We have thus shown that $\log(x) + \log(y) = \log(xy)$. □

Using this property, we can derive two other useful formulas:

$$\log(x^k) = k\log(x),$$

and

$$\log(x) - \log(y) = \log\left(\frac{x}{y}\right).$$

Property 2

This property helps us change from one base to another. We can express the logarithm in any base B in terms of a ratio of logarithms in another base b. The general formula is

$$\log_B(x) = \frac{\log_b(x)}{\log_b(B)}.$$

For example, the logarithm base 10 of a number S can be expressed as a logarithm base 2 or base e as follows:

$$\log_{10}(S) = \frac{\log_{10}(S)}{1} = \frac{\log_{10}(S)}{\log_{10}(10)} = \frac{\log_2(S)}{\log_2(10)} = \frac{\ln(S)}{\ln(10)}.$$

This property is helpful when you need to compute a logarithm in a base that is not available on your calculator. Suppose you're asked to compute $\log_7(S)$, but your calculator only has a $\boxed{\log_{10}}$ button. You can simulate $\log_7(S)$ by computing $\log_{10}(S)$ and dividing by $\log_{10}(7)$.

Exercises

E3.6 Use the properties of logarithms to simplify the expressions

a) $\log(x) + \log(2y)$ **b)** $\log(z) - \log(z^2)$ **c)** $\log(x) + \log(y/x)$

d) $\log_2(8)$ **e)** $\log_3(\frac{1}{27})$ **f)** $\log_{10}(10000)$

Chapter 4

Coordinate systems

Numbers have a dual nature: they're simultaneously abstract mathematical concepts and concrete real-world quantities. The number four corresponds to the abstract concept of "fourness" that is shared by all groups of four objects, but four can also be represented as the length of a four-metre-long wooden pole. This dual nature of numbers is what makes them so useful: we can use numbers and algebra to solve real-world problems, and also visualize abstract algebra problems in terms of real-world lengths. If you want to know the combined length of a four-metre pole and a two-metre pole, you can perform the calculation using numbers inside your head, without needing to move any real poles. Alternatively, you can learn about the abstract concepts of addition and multiplication by manipulating physical objects.[1]

It is possible to add more "dimensions" to the correspondence between numbers and lengths. The Cartesian plane is a two-dimensional *coordinate system* that assigns a pair of coordinates (x,y) to every point in a two-dimensional plane. In this chapter we'll discuss this important visualization tool that will support your understanding of functions (Chapter 5), geometry (Chapter 6), and vectors (Chapter 7).

Many important math ideas can be expressed visually as two-dimensional pictures, geometric shapes, and curves. The same math ideas can also be described by abstract math concepts like numbers and equations. Coordinate systems serve as the bridge between the space you can visualize and the space of abstract math concepts like coordinates and equations.

[1]For example, playing with Cuisenaire rods is an excellent way to become familiar with numbers and math operations.

4.1 The Cartesian plane

The Cartesian plane, named after famous philosopher and mathematician René Descartes, is used to visualize pairs of numbers (x, y).

Recall the number line representation for numbers that we introduced in Section 1.3.

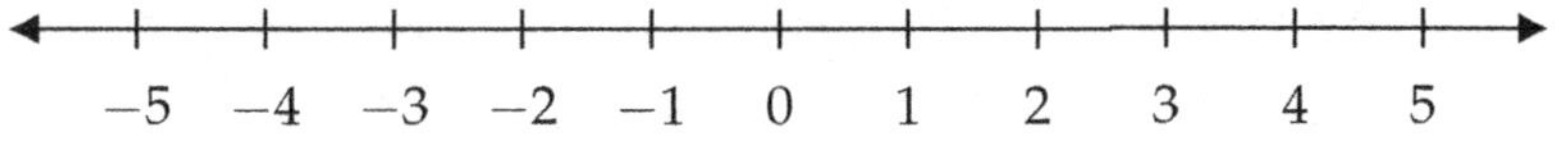

Figure 4.1: Every real number x corresponds to a point on the number line. The number line extends indefinitely to the left (toward negative infinity) and to the right (toward positive infinity).

The Cartesian plane is the two-dimensional generalization of the number line. Generally, we call the plane's horizontal axis "the x-axis" and its vertical axis "the y-axis." We put notches at regular intervals on each axis so we can measure distances.

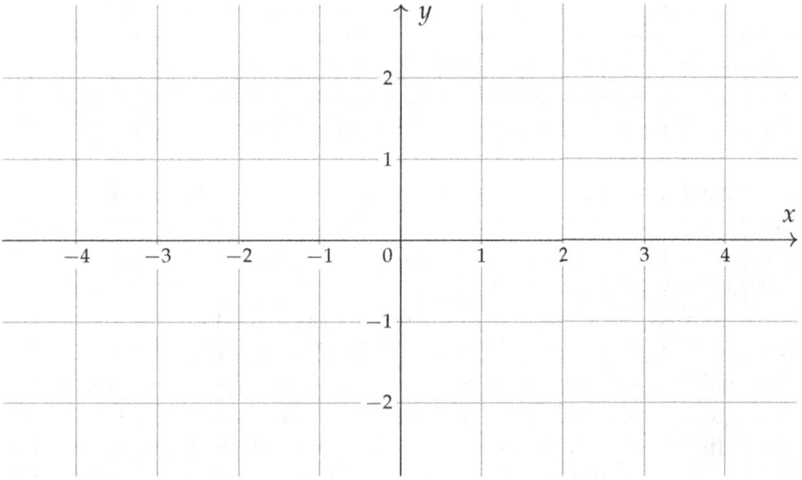

Figure 4.2: Every point in the Cartesian plane corresponds to a pair of real numbers (x, y). Points $P = (P_x, P_y)$, vectors $\vec{v} = (v_x, v_y)$, and graphs of functions $(x, f(x))$ live here.

Figure 4.2 is an example of an empty Cartesian coordinate system. Think of the coordinate system as an empty canvas. What can you draw on this canvas?

Vectors and points

A *point* $P = (P_x, P_y)$ in the Cartesian plane has an x-coordinate and a y-coordinate. To find this point, start from the origin—the point (0,0)—and move a distance P_x on the x-axis, then move a distance P_y on the y-axis.

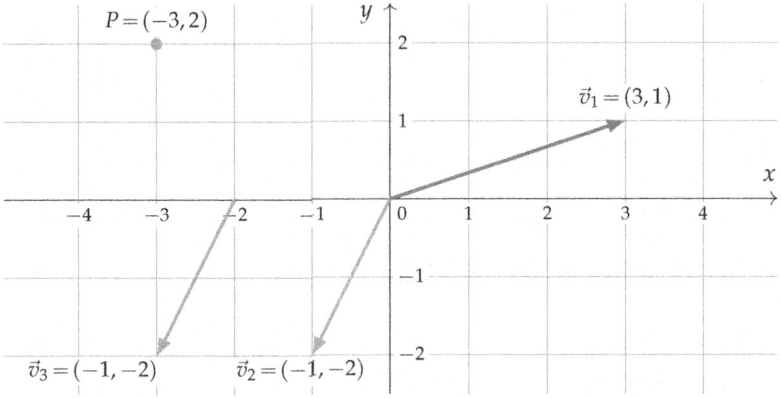

Figure 4.3: A Cartesian plane which shows the point $P = (-3,2)$ and the vectors $\vec{v}_1 = (3,1)$ and $\vec{v}_2 = \vec{v}_3 = (-1,-2)$.

Similar to a point, a vector $\vec{v} = (v_x, v_y)$ is a pair of coordinates. Unlike points, we don't necessarily start from the plane's origin when mapping vectors. We draw vectors as arrows that explicitly mark where the vector starts and where it ends. Note that vectors $\vec{v}_2$ and $\vec{v}_3$ illustrated in Figure 4.3 are actually the *same* vector—the "displace left by 1 and down by 2" vector. It doesn't matter where you draw this vector, it will always be the same whether it begins at the plane's origin or elsewhere. We'll discuss vectors in more details in Chapter 7.

Graphs of functions

The Cartesian plane is great for visualizing functions. You can think of a function as a set of input-output pairs $(x, f(x))$. You can draw the *graph* of a function by letting the y-coordinate represent the function's output value:

$$(x, y) = (x, f(x)).$$

For example, with the function $f(x) = x^2$, we can pass a line through the set of points

$$(x, y) = (x, x^2),$$

and obtain the graph shown in Figure 4.4.

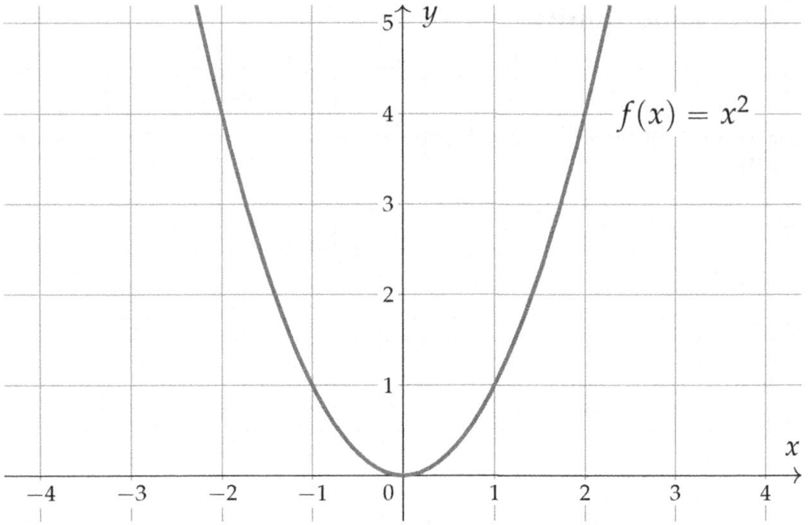

Figure 4.4: The graph of the function $f(x) = x^2$ consists of all pairs of points (x, y) in the Cartesian plane that satisfy $y = x^2$.

When plotting functions by setting $y = f(x)$, we use a special terminology for the two axes. The x-axis represents the *independent* variable (the one that varies freely), and the y-axis represents the *dependent* variable $f(x)$, since $f(x)$ depends on x.

To draw the graph of any function $f(x)$, use the following procedure. Imagine making a sweep over all of the possible input values for the function. For each input x, put a point at the coordinates $(x, y) = (x, f(x))$ in the Cartesian plane. Using the graph of a function, you can literally *see* what the function does: the "height" y of the graph at a given x-coordinate tells you the value of the function $f(x)$.

Dimensions

The number line is one-dimensional. Every number x can be visualized as a point on the number line. The Cartesian plane has two dimensions: the x dimension and the y dimension. If we need to visualize math concepts in 3D, we can use a three-dimensional coordinate system with x, y, and z axes (see Figure 7.10 on page 146).

Chapter 5

Functions

Learning math is like learning a new language. One of the most useful ideas you can learn in the language of math is the concept of a function. If numbers and variables correspond to nouns, functions correspond to verbs. Specifically, functions codify relations between variables; any relationship between two variables can be expressed as a function. This makes functions a powerful tool for modelling real-world situations.

In this chapter we'll learn some important math verbs. We'll start with some general theory about functions, then catalogue the most important functions that occur in everyday life, in scientific modelling, and in business. We're building your vocabulary of math action verbs so you'll be able to get in on the action.

In Section 5.1 we'll introduce the main concepts used to characterize functions: the function's domain and image, the function's graph, and the function's values. Section 5.2 serves as a reference manual for 10 functions of central importance in math, science, and engineering. In Section 5.3 we'll observe how translation and scaling transformations affect the functions' graphs.

5.1 Functions

We need to have a relationship talk. We need to talk about functions. We use functions to describe the relationships between variables. In particular, functions describe how one variable *depends* on another.

For example, the revenue R from a music concert depends on the number of tickets sold n. If each ticket costs \$25, the revenue from the concert can be written *as a function of* n as follows: $R(n) = 25n$. Solving for n in the equation $R(n) = 7000$ tells us the number of ticket sales needed to generate \$7000 in revenue. This is a simple model of a function; as your knowledge of functions builds, you'll learn how to build more detailed models of reality. For instance, if you need to include a 5% processing charge for issuing the tickets, you can update the revenue model to $R(n) = 0.95 \cdot 25 \cdot n$. If the estimated cost of hosting the concert is $C = \$2000$, then the profit from the concert P can be modelled as

$$\begin{aligned} P(n) &= R(n) \ - \ C \\ &= 0.95 \cdot \$25 \cdot n \ - \ \$2000 \end{aligned}$$

The function $P(n) = 23.75n - 2000$ models the profit from the concert as a function of the number of tickets sold. This is a pretty good model already, and you can always update it later as you learn more information.

The more functions you know, the more tools you have for modelling reality. To "know" a function, you must be able to understand and connect several of its aspects. First you need to know the function's mathematical **definition**, which describes exactly what the function does. Starting from the function's definition, you can use your existing math skills to find the function's **properties**. You must also know the **graph** of the function; what the function looks like if you plot x versus $f(x)$ in the Cartesian plane. It's also a good idea to remember the **values** of the function for some important inputs. Finally—and this is the part that takes time—you must learn about the function's **relations** to other functions.

Definitions

A *function* is a mathematical object that takes numbers as inputs and produces numbers as outputs. We use the notation

$$f\colon A \to B$$

to denote a function from the input set A to the output set B. In this book, we mostly study functions that take real numbers as inputs and give real numbers as outputs: $f\colon \mathbb{R} \to \mathbb{R}$.
A function is not a number; rather, it is a *mapping* from numbers to numbers. We say "f maps x to $f(x)$." For any input x, the output value of f for that input is denoted $f(x)$, which is read as "f of x."

We'll now define some fancy technical terms used to describe the input and output sets of functions.

- A: the *source set* of the function describes the types of numbers that the function takes as inputs.
- $\text{Dom}(f)$: the *domain* of a function is the set of allowed input values for the function.

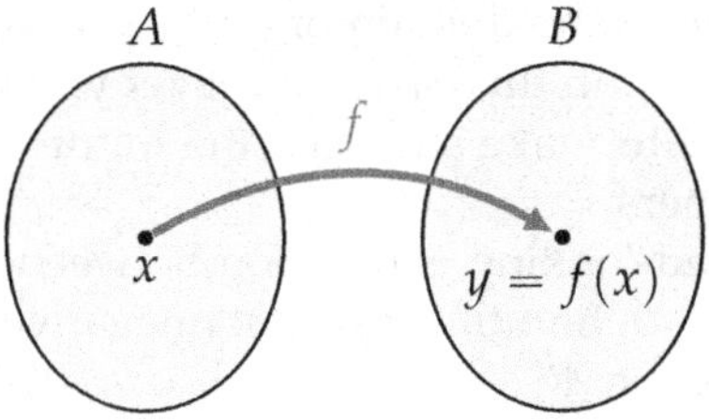

Figure 5.1: An abstract representation of a function f from the set A to the set B. The function f is the arrow which *maps* each input x in A to an output $f(x)$ in B. The output of the function $f(x)$ is also denoted y.

- B: the *target set* of a function describes the type of outputs the function has. The target set is sometimes called the *codomain*.
- $\text{Im}(f)$: the *image* of the function is the set of all possible output values of the function. The image is sometimes called the *range*.

See Figure 5.2 for an illustration of these concepts. The purpose of introducing all this math terminology is so we'll have words to distinguish the general types of inputs and outputs of the function (real numbers, complex numbers, vectors) from the specific properties of the function like its domain and image.

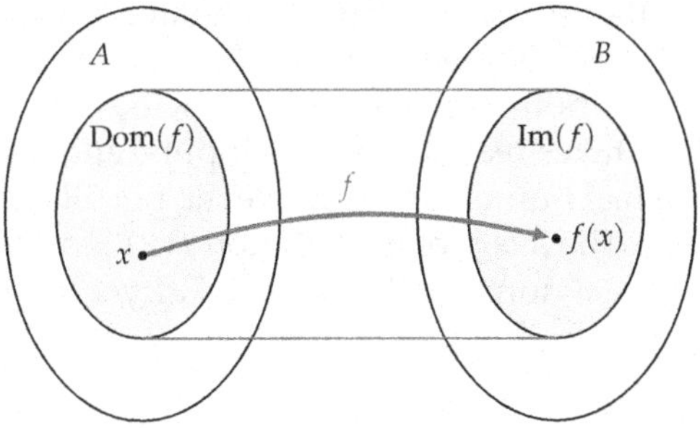

Figure 5.2: Illustration of the input and output sets of a function $f\colon A \to B$. The *source set* is denoted A and the *domain* is denoted $\text{Dom}(f)$. Note that the function's domain is a subset of its source set. The *target set* is denoted B and the *image* is denoted $\text{Im}(f)$. The image is a subset of the target set.

Let's look at an example to illustrate the difference between the source set and the domain of a function. Consider the square root function $f\colon \mathbb{R} \to \mathbb{R}$ defined as $f(x) = \sqrt{x}$, which is shown in Figure 5.3. The source set of f is the set of real numbers—yet only nonnegative real numbers are allowed as inputs, since $\sqrt{x}$ is not defined for negative numbers. Therefore, the domain of the square root function is only the nonnegative real numbers: $\text{Dom}(f) = \mathbb{R}_+ = \{x \in$

$\mathbb{R} \mid x \geqslant 0\}$. Knowing the domain of a function is essential to using the function correctly. In this case, whenever you use the square root function, you need to make sure that the inputs to the function are nonnegative numbers.

The complicated-looking expression between the curly brackets uses *set notation* to define the set of nonnegative numbers $\mathbb{R}_+$. In words, the expression $\mathbb{R}_+ = \{x \in \mathbb{R} \mid x \geqslant 0\}$ states that "$\mathbb{R}_+$ is defined as the set of all real numbers x such that x is greater than or equal to zero." We'll discuss set notation in more detail in Section 8.3. For now, you can just remember that $\mathbb{R}_+$ represents the set of nonnegative real numbers.

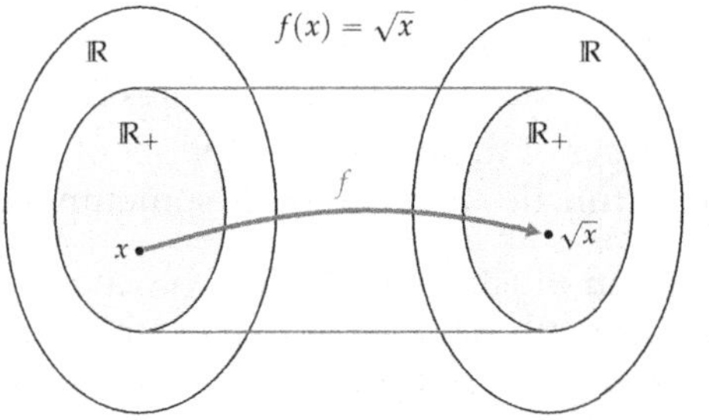

Figure 5.3: The input and output sets of the function $f(x) = \sqrt{x}$. The domain of f is the set of nonnegative real numbers $\mathbb{R}_+$ and its image is $\mathbb{R}_+$.

To illustrate the difference between the image of a function and its target set, let's look at the function $f(x) = x^2$ shown in Figure 5.4. The quadratic function is of the form $f\colon \mathbb{R} \to \mathbb{R}$. The function's source set is $\mathbb{R}$ (it takes real numbers as inputs) and its target set is $\mathbb{R}$ (the outputs are real numbers too); however, not all real numbers are possible outputs. The *image* of the function $f(x) = x^2$ consists only of the nonnegative real numbers $\mathbb{R}_+ = \{y \in \mathbb{R} \mid y \geqslant 0\}$, since $f(x) \geqslant 0$ for all x.

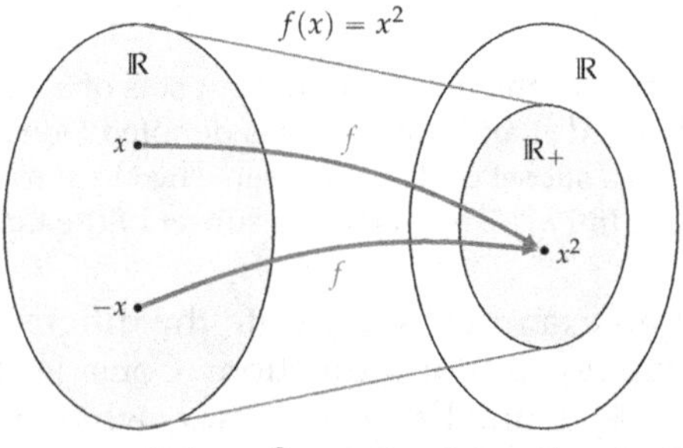

Figure 5.4: The function $f(x) = x^2$ is defined for all reals: $\text{Dom}(f) = \mathbb{R}$. The image of the function is the set of nonnegative real numbers: $\text{Im}(f) = \mathbb{R}_+$.

Function properties

We'll now introduce some additional terminology for describing three important function properties. Every function is a mapping from a source set to a target set, but what kind of mapping is it?

- A function is *injective* if it maps two different inputs to two different outputs. If x_1 and x_2 are two input values that are not equal $x_1 \neq x_2$, then the output values of an injective function will also not be equal $f(x_1) \neq f(x_2)$.
- A function is *surjective* if its image is equal to its target set. For every output y in the target set of a surjective function, there is at least one input x in its domain such that $f(x) = y$.
- A function is *bijective* if it is both injective and surjective.

I know this seems like a lot of terminology to get acquainted with, but it's important to have names for these function properties. We'll need this terminology to give a precise definition of the *inverse function* in the next section.

Injective property We can think of *injective* functions as pipes that transport fluids between containers. Since fluids cannot be compressed, the "output container" must be at least as large as the "input container." If there are two distinct points x_1 and x_2 in the input container of an injective function, then there will be two distinct points $f(x_1)$ and $f(x_2)$ in the output container of the function as well. In other words, injective functions don't smoosh things together.

In contrast, a function that doesn't have the injective property can map several different inputs to the same output value. The function $f(x) = x^2$ is not injective since it sends inputs x and $-x$ to the same output value $f(x) = f(-x) = x^2$, as illustrated in Figure 5.4.

The maps-distinct-inputs-to-distinct-outputs property of injective functions has an important consequence: given the output of an injective function y, there is only one input x such that $f(x) = y$. If a second input x' existed that also leads to the same output $f(x) = f(x') = y$, then the function f wouldn't be injective. For each of the outputs y of an injective function f, there is a *unique* input x such that $f(x) = y$. In other words, injective functions have a unique-input-for-each-output property.

Surjective property A function is *surjective* if its outputs cover the entire target set: every number in the target set is a possible output of the function for some input. For example, the function $f\colon \mathbb{R} \to \mathbb{R}$ defined by $f(x) = x^3$ is surjective: for every number y in the target set $\mathbb{R}$, there is an input x, namely $x = \sqrt[3]{y}$, such that $f(x) = y$. The function $f(x) = x^3$ is surjective since its image is equal to its target set, $\text{Im}(f) = \mathbb{R}$, as shown in Figure 5.5.

On the other hand, the function $f\colon \mathbb{R} \to \mathbb{R}$ defined by the equation $f(x) = x^2$ is not surjective since its image is only the nonnegative numbers $\mathbb{R}_+$ and not the whole set of real numbers (see Figure 5.4). The outputs of this function do not include the negative numbers of the target set, because there is no real number x that can be used as an input to obtain a negative output value.

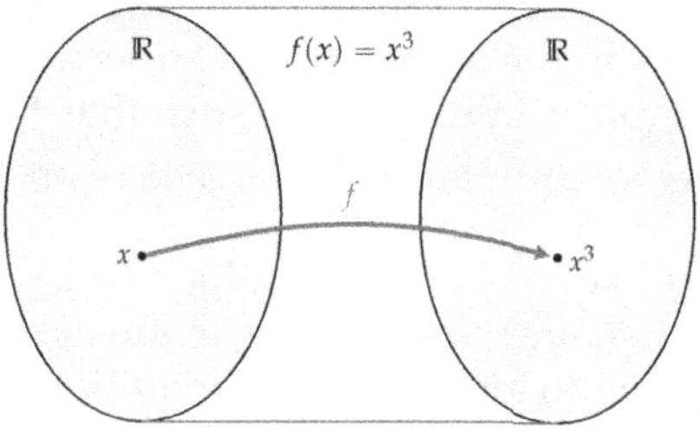

Figure 5.5: For the function $f(x) = x^3$ the image is equal to the target set of the function, $\text{Im}(f) = \mathbb{R}$, therefore the function f is surjective. The function f maps two different inputs $x_1 \neq x_2$ to two different outputs $f(x_1) \neq f(x_2)$, so f is injective. Since f is both injective and surjective, it is a *bijective* function.

Bijective property A function is bijective if it is both injective and surjective. When a function $f : A \to B$ has both the injective and surjective properties, it defines a *one-to-one correspondence* between the numbers of the source set A and the numbers of the target set B. This means for every input value x, there is exactly one corresponding output value y, and for every output value y, there is exactly one input value x such that $f(x) = y$. An example of a bijective function is the function $f\colon \mathbb{R} \to \mathbb{R}$ defined by $f(x) = x^3$ (see Figure 5.5). For every input x in the source set $\mathbb{R}$, the corresponding output y is given by $y = f(x) = x^3$. For every output value y in the target set $\mathbb{R}$, the corresponding input value x is given by $x = \sqrt[3]{y}$.

A function is not bijective if it lacks one of the required properties. Examples of non-bijective functions are $f(x) = \sqrt{x}$, which is not surjective and $f(x) = x^2$, which is neither injective nor surjective.

Counting solutions Another way to understand the injective, surjective, and bijective properties of functions is to think about the solutions to the equation $f(x) = b$, where b is a number in the target set B. The function f is injective if the equation $f(x) = b$ has *at most one* solution for every number b. The function f is surjective if the equation $f(x) = b$ has *at least one* solution for every number b. If the function f is bijective then it is both injective and surjective, which means the equation $f(x) = b$ has *exactly one* solution.

Inverse function

We used inverse functions repeatedly in previous chapters, each time describing the inverse function informally as an "undo" operation. Now that we have learned about bijective functions, we can give a the precise definition of the inverse function and explain some of the details we glossed over previously.

Recall that a *bijective* function $f : A \to B$ is a *one-to-one correspondence* between the numbers in the source set A and numbers in the target set B: for every output y, there is exactly one corresponding input value x such that $f(x) = y$. The *inverse function*, denoted f^{-1}, is the function that takes any output value y in the set B and finds the corresponding input value x that produced it $f^{-1}(y) = x$.

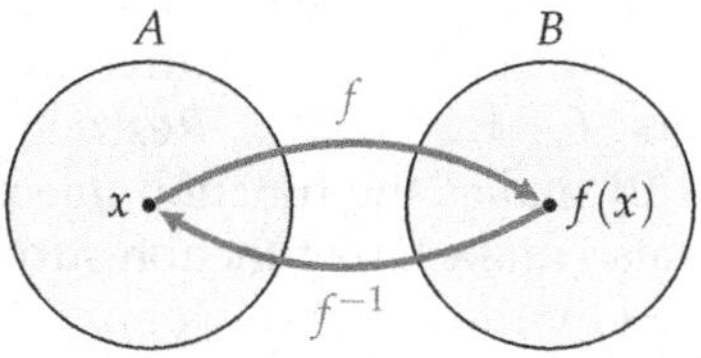

Figure 5.6: The inverse f^{-1} undoes the operation of the function f.

For every bijective function $f : A \to B$, there exists an inverse function $f^{-1} : B \to A$ that performs the *inverse mapping* of f. If we start from some x, apply f, and then apply f^{-1}, we'll arrive—full circle—back to the original input x:

$$f^{-1}(\, f(x)\,) = x.$$

In Figure 5.6 the function f is represented as a forward arrow, and the inverse function f^{-1} is represented as a backward arrow that puts the value $f(x)$ back to the x it came from.

Similarly, we can start from any y in the set B and apply f^{-1} followed by f to get back to the original y we started from:

$$f(\, f^{-1}(y)\,) = y.$$

In words, this equation tells us that f is the "undo" operation for the function f^{-1}, the same way f^{-1} is the "undo" operation for f.

If a function is missing the injective property or the surjective property then it isn't bijective and it doesn't have an inverse. Without the injective property, there could be two inputs x and x' that both produce the same output $f(x) = f(x') = y$. In this case, computing $f^{-1}(y)$ would be impossible since we don't know which of the two possible inputs x or x' was used to produce the output y. Without the surjective property, there could be some output y' in B for which the inverse function f^{-1} is not defined, so the equation $f(f^{-1}(y)) = y$ would not hold for all y in B. The inverse function f^{-1} exists only when the function f is bijective.

Wait a minute! We know the function $f(x) = x^2$ is not bijective and therefore doesn't have an inverse, but we've repeatedly used the square root function as an inverse function for $f(x) = x^2$. What's going on here? Are we using a double standard like a politician that espouses one set of rules publicly, but follows a different set of rules in their private dealings? Is mathematics corrupt?

Don't worry, mathematics is not corrupt—it's all legit. We can use inverses for non-bijective functions by imposing *restrictions* on the source and target sets. The function $f(x) = x^2$ is not bijective when defined as a function $f : \mathbb{R} \to \mathbb{R}$, but it *is* bijective if we define it as a function from the set of nonnegative numbers to the set of nonnegative numbers, $f : \mathbb{R}_+ \to \mathbb{R}_+$. Restricting the source set to $\mathbb{R}_+ = \{x \in \mathbb{R} \mid x \geqslant 0\}$ makes the function injective, and restricting the target set to $\mathbb{R}_+$ also makes the function surjective. The function $f : \mathbb{R}_+ \to \mathbb{R}_+$ defined by the equation $f(x) = x^2$ is bijective and its inverse is $f^{-1}(y) = \sqrt{y}$.

It's important to keep track the of restrictions on the source set we applied when solving equations. For example, solving the equation $x^2 = c$ by restricting the solution space to nonnegative numbers will give us only the positive solution $x = \sqrt{c}$. We have to manually add the negative solution $x = -\sqrt{c}$ in order to obtain the complete solutions: $x = \sqrt{c}$ or $x = -\sqrt{c}$, which is usually written $x = \pm\sqrt{c}$. The possibility of multiple solutions is present whenever we solve equations involving non-injective functions.

Function composition

We can combine two simple functions by chaining them together to build a more complicated function. This act of applying one function after another is called *function composition*. Consider for example the composition:

$$f \circ g\,(x) = f(\,g(x)\,) = z.$$

Figure 5.7 illustrates the concept of function composition. First, the function $g : A \to B$ acts on some input x to produce an intermediary value $y = g(x)$ in the set B. The intermediary value y is then passed

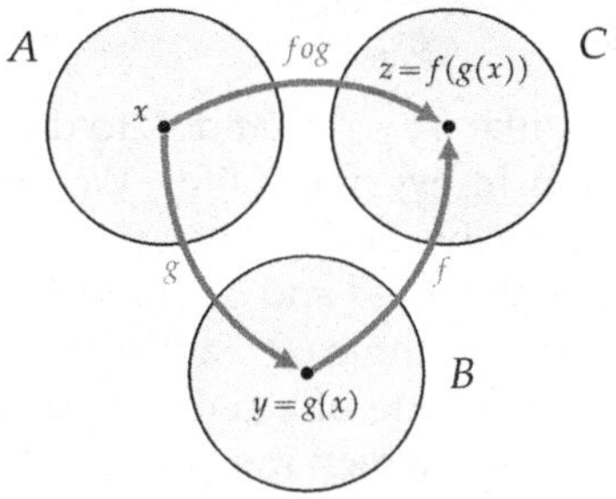

Figure 5.7: The function composition $f \circ g$ describes the combination of first applying the function g, followed by the function f: $f \circ g\,(x) = f(g(x))$.

through the function $f : B \to C$ to produce the final output value $z = f(y) = f(g(x))$ in the set C. We can think of the *composite function* $f \circ g$ as a function in its own right. The function $f \circ g : A \to C$ is defined through the formula $f \circ g\,(x) = f(g(x))$.

Don't worry too much about the "$\circ$" symbol—it's just a convenient math notation I wanted you to know about. Writing $f \circ g$ is the same as writing $f(g(x))$. The important takeaway from Figure 5.7 is that functions can be combined by using the outputs of one function as the inputs to the next. This is a very useful idea for building math models. You can understand many complicated input-output transformations by describing them as compositions of simple functions.

Example 1 Consider the function $g\colon \mathbb{R}_+ \to \mathbb{R}_+$ given by $g(x) = \sqrt{x}$, and the function $f\colon \mathbb{R} \to \mathbb{R}_+$ defined by $f(x) = x^2$. The composite function $f \circ g\,(x) = (\sqrt{x})^2 = x$ is defined for all nonnegative reals. The composite function $g \circ f$ is defined for all real numbers, and we have $g \circ f\,(x) = \sqrt{x^2} = |x|$.

Example 2 The composite functions $f \circ g$ and $g \circ f$ describe different operations. If $g(x) = \ln(x)$ and $f(x) = x^2$, the functions $g \circ f\,(x) = \ln(x^2)$ and $f \circ g\,(x) = (\ln x)^2$ have different domains and produce different outputs, as you can verify using a calculator.

Using the notation "$\circ$" for function composition, we can give a concise description of the properties of a bijective function $f : A \to B$ and its inverse function $f^{-1} : B \to A$:

$$(f^{-1} \circ f)(x) = x \qquad \text{and} \qquad (f \circ f^{-1})(y) = y,$$

for all x in A and all y in B.

Function names

We use short symbols like $+$, $-$, $\times$, and $\div$ to denote most of the important functions used in everyday life. We also use the squiggle notation $\sqrt{\ }$ for square roots and superscripts to denote exponents. All other functions are identified and denoted by their *name*. If I want to compute the *cosine* of the angle $60°$ (a function describing the ratio between the length of one side of a right-angle triangle and the hypotenuse), I write $\cos(60°)$, which means I want the value of the cos function for the input $60°$.

Incidentally, the function cos has a nice output value for that specific angle: $\cos(60°) = \frac{1}{2}$. Therefore, seeing $\cos(60°)$ somewhere in an equation is the same as seeing $\frac{1}{2}$. To find other values of the function, say $\cos(33.13°)$, you'll need a calculator. All scientific calculators have a convenient little [cos] button for this very purpose.

Handles on functions

When you learn about functions you learn about the different "handles" by which you can "grab" these mathematical objects. The main handle for a function is its **definition**: it tells you the precise way to calculate the output when you know the input. The function definition is an important handle, but it is also important to "feel" what the function does intuitively. How does one get a feel for a function?

Table of values

One simple way to represent a function is to look at a list of input-output pairs: $\{\{\text{in} = x_1, \text{out} = f(x_1)\}, \{\text{in} = x_2, \text{out} = f(x_2)\}, \{\text{in} = x_3, \text{out} = f(x_3)\}, \ldots\}$. A more compact notation for the input-output pairs is $\{(x_1, f(x_1)), (x_2, f(x_2)), (x_3, f(x_3)), \ldots\}$, where the first number of each pair represents an input value and the second represents the output value given by the function.

We can also build a **table of values** by writing the input values in one column and recording the corresponding output values in a second column. You can choose inputs at random or focus on the important-looking x values in the function's domain.

You can create a table of values for any function you want to study. Follow the example shown in Table 5.1. Use the input values that interest you and fill out the right side of the table by calculating the value of $f(x)$ for each input x.

input = x	$\rightarrow$	$f(x)$ = output
0	$\rightarrow$	$f(0)$
1	$\rightarrow$	$f(1)$
55	$\rightarrow$	$f(55)$
x_4	$\rightarrow$	$f(x_4)$

Table 5.1: Table of input-output values of the function $f(x)$. The input values $x = 0$, $x = 1$ and $x = 55$ are chosen to "test" what the function does.

Function graph

One of the best ways to feel a function is to look at its graph. A graph is a line on a piece of paper that passes through all input-output pairs of a function. Imagine you have a piece of paper, and on it you draw a blank *coordinate system* as in Figure 5.8.

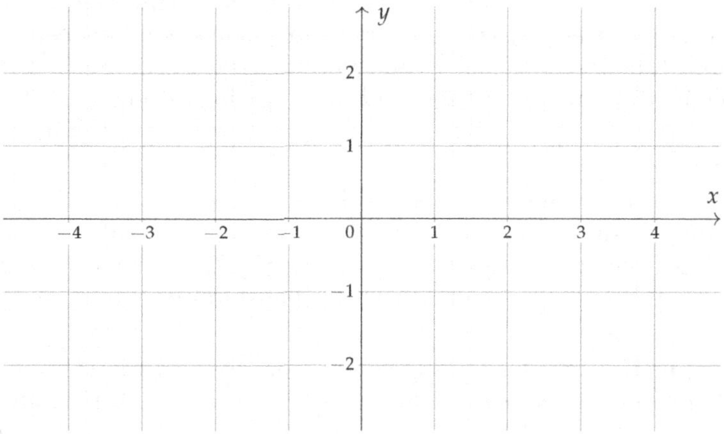

Figure 5.8: An empty (x, y)-coordinate system that you can use to draw function graphs. The graph of $f(x)$ consists of all the points for which $(x, y) = (x, f(x))$. See Figure 4.4 on page 56 for the graph of $f(x) = x^2$.

The horizontal axis is used to measure x. The vertical axis is used to measure $f(x)$. Because writing out $f(x)$ every time is long and tedious, we use a short, single-letter alias to denote the output value of f as follows:

$$y \equiv f(x) = \text{output}.$$

Think of each input-output pair of the function f as a point (x, y) in the coordinate system. The graph of a function is a representational drawing of everything the function does. If you understand how to interpret this drawing, you can infer everything there is to know about the function.

Facts and properties

Another way to feel a function is by knowing the function's properties. This approach boils down to learning facts about the function and its connections to other functions. An example of a mathematical connection is the equation $\log_B(x) = \frac{\log_b(x)}{\log_b(B)}$, which describes a link between the logarithmic function base B and the logarithmic function base b.

The more you know about a function, the more "paths" your brain builds to connect to that function. Real math knowledge is not about memorization; it is about establishing a network of associations between different areas of information in your brain. See the concept maps on page iii for an illustration of the paths that link math concepts. Mathematical thought is the usage of these associations to carry out calculations and produce mathematical arguments. For example, knowing about the connection between logarithmic functions will allow you compute the value of $\log_7(e^3)$, even though calculators don't have a button for logarithms base 7. We find $\log_7(e^3) = \frac{\ln e^3}{\ln 7} = \frac{3}{\ln 7}$, which can be computed using the ln button.

To develop mathematical skills, it is vital to practice path-building between concepts by solving exercises. With this book, I will introduce you to some of the many paths linking math concepts, but it's on you to reinforce these paths through practice.

Example 3 Consider the function f from the real numbers to the real numbers ($f\colon \mathbb{R} \to \mathbb{R}$) defined as $f(x) = x^2 + 2x - 3$. The value of f when $x = 1$ is $f(1) = 1^2 + 2(1) - 3 = 0$. When $x = 2$, the output is $f(2) = 2^2 + 2(2) - 3 = 5$. What is the value of f when $x = 0$? You can use algebra to rewrite this function as $f(x) = (x+3)(x-1)$, which tells you the graph of this function crosses the x-axis at $x = -3$ and at $x = 1$. The values above will help you plot the graph of $f(x)$.

Example 4 Consider the exponential function with base 2 defined by $f(x) = 2^x$. This function is crucial to computer systems. For instance, RAM memory chips come in powers of two because the memory space is exponential in the number of "address lines" used on the chip. When $x = 1$, $f(1) = 2^1 = 2$. When x is 2 we have $f(2) = 2^2 = 4$. The function is therefore described by the following input-output pairs: $(0,1)$, $(1,2)$, $(2,4)$, $(3,8)$, $(4,16)$, $(5,32)$, $(6,64)$, $(7,128)$, $(8,256)$, $(9,512)$, $(10,1024)$, $(11,2048)$, $(12,4096)$, etc. Recall that any number raised to exponent 0 gives 1. Thus, the exponential function passes through the point $(0,1)$. Recall also that negative exponents lead to fractions, so we have the points $(-1,\frac{1}{2})$, $(-2,\frac{1}{4})$, $(-3,\frac{1}{8})$, etc. You can plot these $(x, f(x))$ coordinates in the Cartesian plane to obtain the graph of the function.

Discussion

To describe a function we specify its source and target sets $f\colon A \to B$, then give an equation of the form $f(x) =$ "expression involving x" that defines the function. Since functions are defined using equations, does this mean that functions and equations are the same thing? Let's take a closer look.

In general, any equation containing two variables describes a *relation* between these variables. For example, the equation $x-3 = y-4$ describes a relation between the variables x and y. We can isolate the variable y in this equation to obtain $y = x+1$ and thus find the value of y when the value of x is given. We can also isolate x to obtain $x = y-1$ and use this equation to find x when the value of y is given. In the context of an equation, the relationship between the variables x and y is symmetrical and no special significance is attached to either of the two variables.

We also can describe the same relationship between x and y as a function $f : \mathbb{R} \to \mathbb{R}$. We choose to identify x as the input variable and y as the output variable of the function f. Having identified y with the output variable, we can interpret the equation $y = x+1$ as the definition of the function $f(x) = x+1$.

Note that the equation $x-3 = y-4$ and the function $f(x) = x+1$ describe the same relationship between the variables x and y. For example, if we set the value $x = 5$ we can find the value of y by solving the equation $5-3 = y-4$ to obtain $y = 6$, or by computing the output of the function $f(x)$ for the input $x = 5$, which gives us the same answer $f(5) = 6$. In both cases we arrive at the same answer, but modelling the relationship between x and y as a function allows us to use the whole functions toolbox, like function composition and function inverses.

In this section we talked a lot about functions in general but we haven't said much about any function specifically. There are many useful functions out there, and we can't discuss them all here. In the next section, we'll introduce 10 functions of strategic importance for all of science. If you get a grip on these functions, you'll be able to understand all of physics and calculus and handle *any* problem your teacher may throw at you.

5.2 Functions reference

Your *function vocabulary* determines how well you can express yourself mathematically in the same way your English vocabulary determines how well you can express yourself in English. The following pages aim to embiggen your function vocabulary, so you'll know how to handle the situation when a teacher tries to pull some trick on you at the final. Here are the ten most important functions in math:

1. Straight line $f(x) = mx + b$ (see pages 71 and 165)
2. Quadratic function $f(x) = x^2$ (pages 73, 90, and 120)
3. Square root $f(x) = \sqrt{x}$ (page 74)
4. Absolute value $f(x) = |x|$ (page 75)
5. Polynomials $f(x) = a_0 + a_1x + a_2x^2 + \cdots + a_nx^n$ (page 76)
6. Sine $f(x) = \sin(x)$ (pages 80, 91, and 99)
7. Cosine $f(x) = \cos(x)$ (pages 82 and 99)
8. Tangent $f(x) = \tan(x)$ (page 83)
9. Exponential $f(x) = e^x$ (pages 84 and 170)
10. Logarithm $f(x) = \ln(x)$ (page 85)

If you're seeing these functions for the first time, don't worry about remembering all the facts and properties on the first reading. We'll use these functions throughout the rest of the book, so you'll have plenty of time to become familiar with them. Remember to return to this section if you ever get stuck on a function.

To build mathematical intuition, it's essential you understand functions' graphs. Memorizing the definitions and properties of functions gets a lot easier with visual accompaniment. Indeed, remembering what the function "looks like" is a great way to train yourself to recognize various types of functions. Figure 5.9 shows the graphs of some of the functions we'll use in this book.

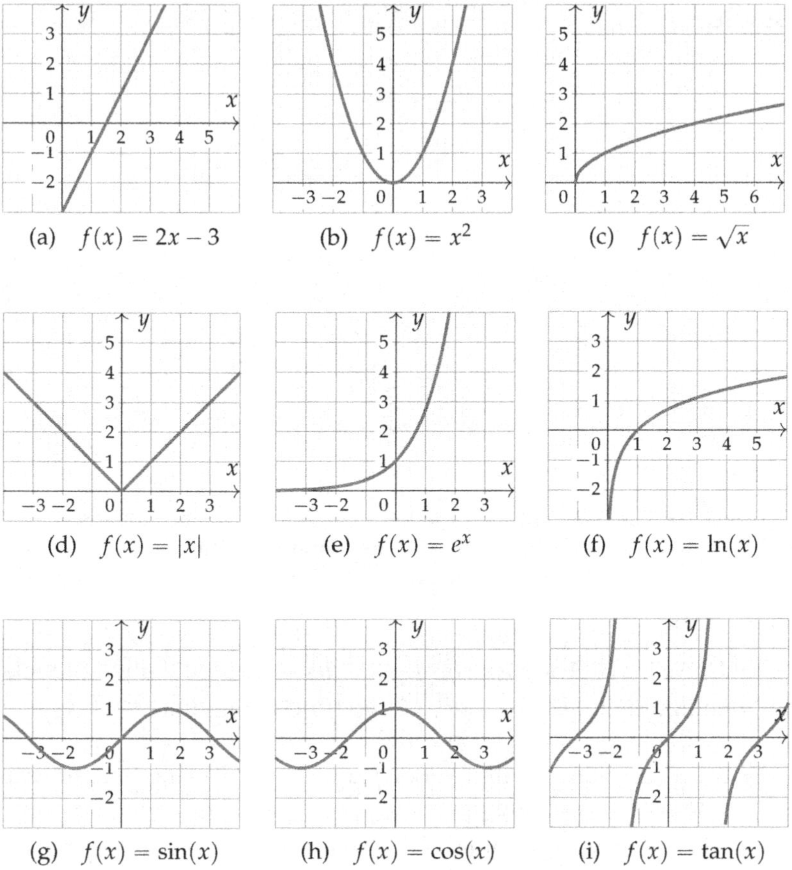

(a) $f(x) = 2x - 3$ (b) $f(x) = x^2$ (c) $f(x) = \sqrt{x}$

(d) $f(x) = |x|$ (e) $f(x) = e^x$ (f) $f(x) = \ln(x)$

(g) $f(x) = \sin(x)$ (h) $f(x) = \cos(x)$ (i) $f(x) = \tan(x)$

Figure 5.9: We'll see many types of function graphs in the next pages.

Line

The equation of a line describes an input-output relationship where the change in the output is *proportional* to the change in the input. The equation of a line is

$$f(x) = mx + b.$$

The constant m describes the slope of the line. The constant b is called the y-intercept and it is the value of the function when $x = 0$.

Consider what relationship the equation of $f(x)$ describes for different values of m and b. What happens when m is positive? What happens when m is negative?

Graph

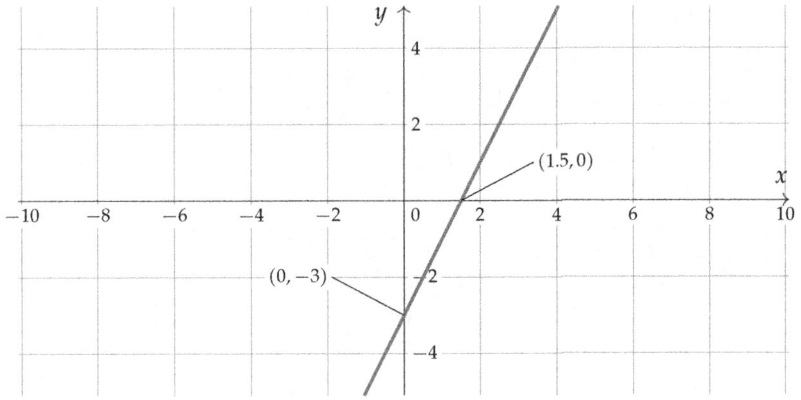

Figure 5.10: The graph of the function $f(x) = 2x - 3$. The slope is $m = 2$. The y-intercept of this line is $b = -3$. The x-intercept is at $x = \frac{3}{2}$.

Properties

- Domain: $\mathbb{R}$. The function $f(x) = mx + b$ is defined for all reals.
- Image: $\mathbb{R}$ if $m \neq 0$. If $m = 0$ the function is constant $f(x) = b$, so the image set contains only a single number $\{b\}$.
- $x = -b/m$: the x-intercept of $f(x) = mx + b$. The x-intercept is obtained by solving $f(x) = 0$.
- The inverse to the line $f(x) = mx + b$ is $f^{-1}(x) = \frac{1}{m}(x - b)$, which is also a line.

General equation

A line can also be described in a more symmetric form as a relation:

$$Ax + By = C.$$

This is known as the *general* equation of a line. The general equation for the line shown in Figure 5.10 is $2x - 1y = 3$.

Given the general equation of a line $Ax + By = C$ with $B \neq 0$, you can convert to the function form $y = f(x) = mx + b$ by computing the slope $m = \frac{-A}{B}$ and the y-intercept $b = \frac{C}{B}$.

Square

The function x *squared*, is also called the *quadratic* function, or *parabola*. The formula for the quadratic function is

$$f(x) = x^2.$$

The name "quadratic" comes from the Latin *quadratus* for square, since the expression for the area of a square with side length x is x^2.

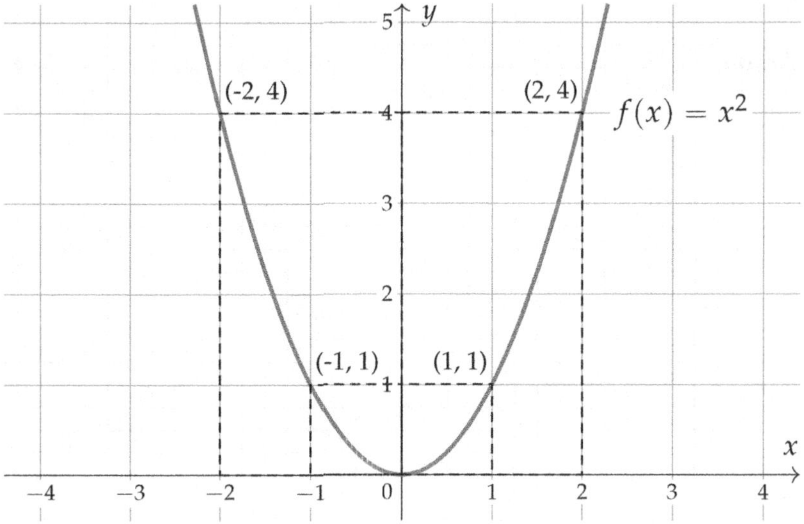

Figure 5.11: Plot of the quadratic function $f(x) = x^2$. The graph of the function passes through the following (x, y) coordinates: $(-2, 4)$, $(-1, 1)$, $(0, 0)$, $(1, 1)$, $(2, 4)$, $(3, 9)$, etc.

Properties

- Domain: $\mathbb{R}$. The function $f(x) = x^2$ is defined for all numbers.
- Image: $\mathbb{R}_+ = \{y \in \mathbb{R} \mid y \geqslant 0\}$. The outputs are nonnegative numbers since $x^2 \geqslant 0$, for all real numbers x.
- The function x^2 is the inverse of the square root function $\sqrt{x}$.
- $f(x) = x^2$ is *two-to-one*: it sends both x and $-x$ to the same output value $x^2 = (-x)^2$.
- The quadratic function is *convex*, meaning it curves upward.

The set expression $\{y \in \mathbb{R} \mid y \geqslant 0\}$ that we use to define the nonnegative real numbers ($\mathbb{R}_+$) is read "the set of real numbers that are greater than or equal to zero."

Square root

The square root function is denoted

$$f(x) = \sqrt{x} = x^{\frac{1}{2}}.$$

The square root $\sqrt{x}$ is the inverse function of the square function x^2 when the two functions are defined as $f : \mathbb{R}_+ \to \mathbb{R}_+$. The symbol $\sqrt{c}$ refers to the *positive* solution of $x^2 = c$. Note that $-\sqrt{c}$ is also a solution of $x^2 = c$.

Graph

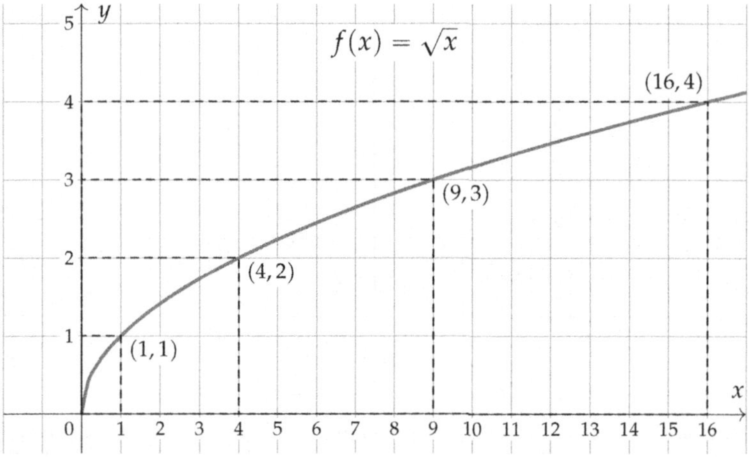

Figure 5.12: The graph of the function $f(x) = \sqrt{x}$. The domain of the function is $\mathbb{R}_+$ because we can't take the square root of a negative number.

Properties

- Domain: $\mathbb{R}_+ = \{x \in \mathbb{R} \mid x \geqslant 0\}$. The function $f(x) = \sqrt{x}$ is only defined for nonnegative inputs. There is no real number y such that y^2 is negative, hence the function $f(x) = \sqrt{x}$ is not defined for negative inputs x.
- Image: $\mathbb{R}_+ = \{y \in \mathbb{R} \mid y \geqslant 0\}$. The outputs of the function $f(x) = \sqrt{x}$ are nonnegative numbers since $\sqrt{x} \geqslant 0$.

In addition to *square* root, there is also *cube* root $f(x) = \sqrt[3]{x} = x^{\frac{1}{3}}$, which is the inverse function for the cubic function $f(x) = x^3$. We have $\sqrt[3]{8} = 2$ since $2 \times 2 \times 2 = 8$. More generally, we can define the nth-root function $\sqrt[n]{x}$ as the inverse function of x^n.

Absolute value

The *absolute value* function tells us the size of numbers without paying attention to whether the number is positive or negative. We can compute a number's absolute value by *ignoring the sign* of the number. A number's absolute value corresponds to its distance from the origin of the number line.

Another way of thinking about the absolute value function is to say it multiplies negative numbers by -1 to "cancel" their negative sign:

$$f(x) = |x| = \begin{cases} x & \text{if } x \geqslant 0, \\ -x & \text{if } x < 0. \end{cases}$$

Graph

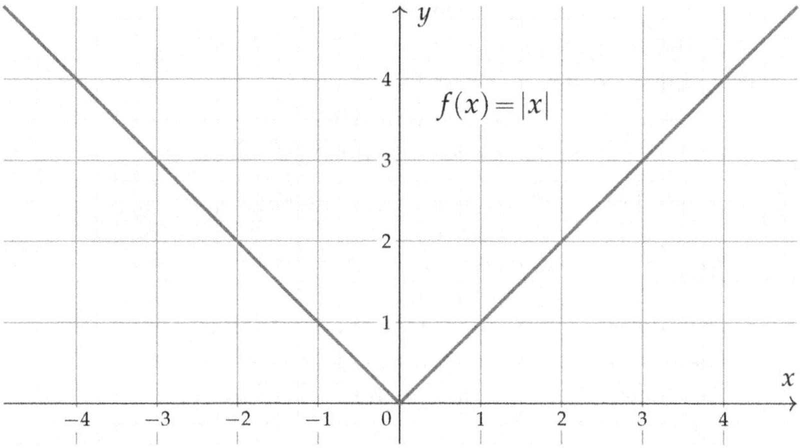

Figure 5.13: The graph of the absolute value function $f(x) = |x|$.

Properties

- Domain: $\mathbb{R}$. The function $f(x) = |x|$ is defined for all inputs.
- Image: $\mathbb{R}_+ = \{y \in \mathbb{R} \mid y \geqslant 0\}$
- The combination of squaring followed by square-root is equivalent to the absolute value function:

$$\sqrt{x^2} = |x|,$$

since squaring destroys the sign.

Polynomials

The polynomials are a very useful family of functions. For example, quadratic polynomials of the form $f(x) = ax^2 + bx + c$ often arise when describing physics phenomena.

The general equation for a polynomial function of degree n is

$$f(x) = a_0 + a_1x + a_2x^2 + a_3x^3 + \cdots + a_nx^n.$$

The constants a_i are known as the *coefficients* of the polynomial.

Parameters

- x: the variable
- a_0: the constant term
- a_1: the *linear* coefficient, or *first-order* coefficient
- a_2: the *quadratic* coefficient
- a_3: the *cubic* coefficient
- a_n: the nth order coefficient
- n: the *degree* of the polynomial. The degree of $f(x)$ is the largest power of x that appears in the polynomial.

A polynomial of degree n has $n+1$ coefficients: $a_0, a_1, a_2, \ldots, a_n$.

Properties

- Domain: $\mathbb{R}$. Polynomials are defined for all inputs.
- The roots of $f(x)$ are the values of x for which $f(x) = 0$.
- The image of a polynomial function depends on the coefficients.
- The sum of two polynomials is also a polynomial.

The most general first-degree polynomial is a line $f(x) = mx + b$, where m and b are arbitrary constants. The most general second-degree polynomial is $f(x) = a_2x^2 + a_1x + a_0$, where again a_0, a_1, and a_2 are arbitrary constants. We call a_k the *coefficient* of x^k, since this is the number that appears in front of x^k. Following the pattern, a third-degree polynomial will look like $f(x) = a_3x^3 + a_2x^2 + a_1x + a_0$.

In general, a polynomial of degree n has the equation

$$f(x) = a_nx^n + a_{n-1}x^{n-1} + \cdots + a_2x^2 + a_1x + a_0.$$

You can add two polynomials by adding together their coefficients:

$$\begin{aligned} f(x) + g(x) &= (a_nx^n + \cdots + a_1x + a_0) + (b_nx^n + \cdots + b_1x + b_0) \\ &= (a_n + b_n)x^n + \cdots + (a_1 + b_1)x + (a_0 + b_0). \end{aligned}$$

The subtraction of two polynomials works similarly. We can also multiply polynomials together using the general algebra rules for expanding brackets.

Solving polynomial equations

Very often in math, you will have to *solve* polynomial equations of the form

$$A(x) = B(x),$$

where $A(x)$ and $B(x)$ are both polynomials. Recall from earlier that to *solve*, we must find the values of x that make the equality true.

Say the revenue of your company is a function of the number of products sold x, and can be expressed as $R(x) = 2x^2 + 2x$. Say also the cost you incur to produce x objects is $C(x) = x^2 + 5x + 10$. You want to determine the amount of product you need to produce to break even, that is, so that revenue equals cost: $R(x) = C(x)$. To find the break-even value x, solve the equation

$$2x^2 + 2x = x^2 + 5x + 10.$$

This may seem complicated since there are xs all over the place. No worries! We can turn the equation into its "standard form," and then use the quadratic formula. First, move all the terms to one side until only zero remains on the other side:

$$\begin{aligned}
2x^2 + 2x \quad - x^2 &= \cancel{x^2} + 5x + 10 \quad - \cancel{x^2} \\
x^2 + 2x \quad - 5x &= \cancel{5x} + 10 \quad - \cancel{5x} \\
x^2 - 3x \quad - 10 &= \cancel{10} \quad - \cancel{10} \\
x^2 - 3x - 10 &= 0.
\end{aligned}$$

Remember, if we perform the same operations on both sides of the equation, the resulting equation has the same solutions. Therefore, the values of x that satisfy $x^2 - 3x - 10 = 0$, namely $x = -2$ and $x = 5$, also satisfy $2x^2 + 2x = x^2 + 5x + 10$, which is the original problem we're trying to solve.

This "shuffling of terms" approach will work for any polynomial equation $A(x) = B(x)$. We can always rewrite it as $C(x) = 0$, where $C(x)$ is a new polynomial with coefficients equal to the difference of the coefficients of A and B. Don't worry about which side you move all the coefficients to because $C(x) = 0$ and $0 = -C(x)$ have exactly the same solutions. Furthermore, the degree of the polynomial C can be no greater than that of A or B.

The form $C(x) = 0$ is the *standard form* of a polynomial, and we'll explore several formulas you can use to find its solution(s).

Formulas

The formula for solving the polynomial equation $P(x) = 0$ depends on the *degree* of the polynomial in question.

For a first-degree polynomial equation, $P_1(x) = mx + b = 0$, the solution is $x = \frac{-b}{m}$: just move b to the other side and divide by m.

For a second-degree polynomial,

$$P_2(x) = ax^2 + bx + c = 0,$$

the solutions are $x_1 = \frac{-b+\sqrt{b^2-4ac}}{2a}$ and $x_2 = \frac{-b-\sqrt{b^2-4ac}}{2a}$.

If $b^2 - 4ac < 0$, the solutions will involve taking the square root of a negative number. In those cases, we say no real solutions exist.

There is also a formula for polynomials of degree 3 and 4, but they are complicated. For polynomials with order $\geqslant 5$, there does not exist a general analytical solution.

Using a computer

When solving real-world problems, you'll often run into much more complicated equations. To find the solutions of anything more complicated than the quadratic equation, I recommend using a computer algebra system like SymPy: http://live.sympy.org.

To make SymPy solve the standard-form equation $C(x) = 0$, call the function `solve(expr,var)`, where the expression `expr` corresponds to $C(x)$, and `var` is the variable you want to solve for. For example, to solve $x^2 - 3x + 2 = 0$, type in the following:

```
>>> solve(x**2 - 3*x + 2, x)            # usage: solve(expr, var)
[1, 2]
```

The function `solve` will find the solutions to any equation of the form `expr = 0`. In this case, we see the solutions are $x = 1$ and $x = 2$.

Another way to solve the equation is to factor the polynomial $C(x)$ using the function `factor` like this:

```
>>> factor(x**2 - 3*x + 2)              # usage: factor(expr)
(x - 1)*(x - 2)
```

We see that $x^2 - 3x + 2 = (x - 1)(x - 2)$, which confirms the two roots are indeed $x = 1$ and $x = 2$.

To learn more about SymPy, check out Appendix C on page 217, which talks about all the SymPy functions that are available to you.

Substitution trick

Sometimes you can solve fourth-degree polynomials by using the quadratic formula. Say you're asked to solve for x in

$$x^4 - 7x^2 + 10 = 0.$$

Imagine this problem is on your exam, where you are not allowed to use a computer. How does the teacher expect you to solve for x? The trick is to substitute $y = x^2$ and rewrite the same equation as

$$y^2 - 7y + 10 = 0,$$

which you can solve by applying the quadratic formula. If you obtain the solutions $y = \alpha$ and $y = \beta$, then the solutions to the original fourth-degree polynomial are $x = \pm\sqrt{\alpha}$ and $x = \pm\sqrt{\beta}$, since $y = x^2$.

Since we're not taking an exam right now, we are allowed to use the computer to find the roots:

```
>>> solve(y**2 - 7*y + 10, y)
[2, 5]
>>> solve(x**4 - 7*x**2 + 10, x)
[sqrt(2), -sqrt(2), sqrt(5), -sqrt(5)]
```

Note how the second-degree polynomial has two roots, while the fourth-degree polynomial has four roots.

Even and odd functions

The polynomials form an entire family of functions. Depending on the choice of degree n and coefficients a_0, a_1, ..., a_n, a polynomial function can take on many different shapes. Consider the following observations about the symmetries of polynomials:

- If a polynomial contains only even powers of x, like $f(x) = 1 + x^2 - x^4$ for example, we call this polynomial *even*. Even polynomials have the property $f(x) = f(-x)$. The sign of the input doesn't matter.
- If a polynomial contains only odd powers of x, for example $g(x) = x + x^3 - x^9$, we call this polynomial *odd*. Odd polynomials have the property $g(x) = -g(-x)$.
- If a polynomial has both even and odd terms then it is neither even nor odd.

The terminology of *odd* and *even* applies to functions in general and not just to polynomials. All functions that satisfy $f(x) = f(-x)$ are called *even functions*, and all functions that satisfy $f(x) = -f(-x)$ are called *odd functions*.

Sine

The sine function represents a fundamental unit of vibration. The graph of $\sin(x)$ *oscillates* up and down and crosses the x-axis multiple times. The shape of the graph of $\sin(x)$ corresponds to the shape of a vibrating string. See Figure 5.14.

In the remainder of this book, we'll meet the function $\sin(x)$ many times. We'll define the function $\sin(x)$ more formally as a trigonometric ratio in Section 6.2. In Chapter 7 we'll use $\sin(x)$ and $\cos(x)$ (another trigonometric ratio) to work out the *components* of vectors. The sine function also describes waves and periodic motion.

At this point in the book, however, we don't want to go into too much detail about all these applications. Let's hold off on the discussion about vectors, triangles, angles, and ratios of lengths of sides and instead just focus on the graph of the function $f(x) = \sin(x)$.

Graph

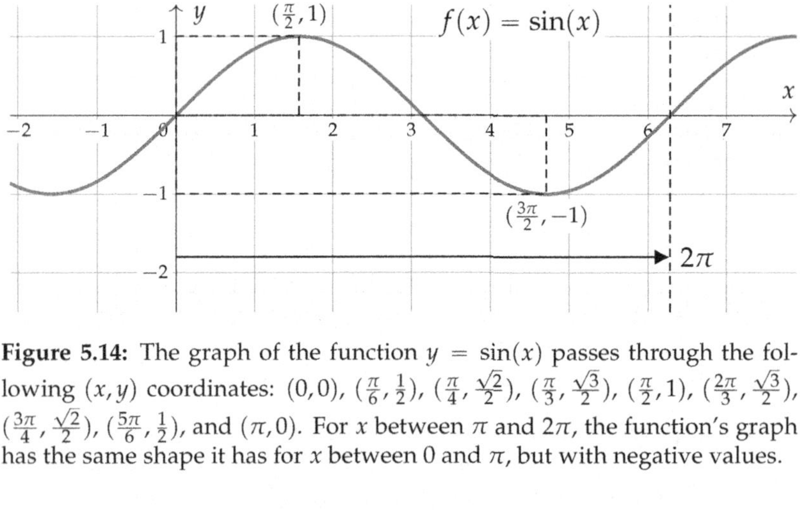

Figure 5.14: The graph of the function $y = \sin(x)$ passes through the following (x,y) coordinates: $(0,0)$, $(\frac{\pi}{6},\frac{1}{2})$, $(\frac{\pi}{4},\frac{\sqrt{2}}{2})$, $(\frac{\pi}{3},\frac{\sqrt{3}}{2})$, $(\frac{\pi}{2},1)$, $(\frac{2\pi}{3},\frac{\sqrt{3}}{2})$, $(\frac{3\pi}{4},\frac{\sqrt{2}}{2})$, $(\frac{5\pi}{6},\frac{1}{2})$, and $(\pi,0)$. For x between π and 2π, the function's graph has the same shape it has for x between 0 and π, but with negative values.

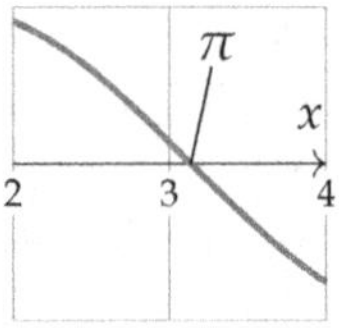

Figure 5.15: The function $f(x) = \sin(x)$ crosses the x-axis at $x = \pi$.

Let's start at $x = 0$ and follow the graph of the function $\sin(x)$ as it goes up and down. The graph starts from $(0,0)$ and smoothly increases until it reaches the maximum value at $x = \frac{\pi}{2}$. Afterward, the function comes back down to cross the x-axis at $x = \pi$. After π, the function drops below the x-axis and reaches its minimum value of -1 at $x = \frac{3\pi}{2}$. It then travels up again to cross the x-axis at $x = 2\pi$. This 2π-long cycle repeats after $x = 2\pi$. This is why we call the function *periodic*—the shape of the graph repeats.

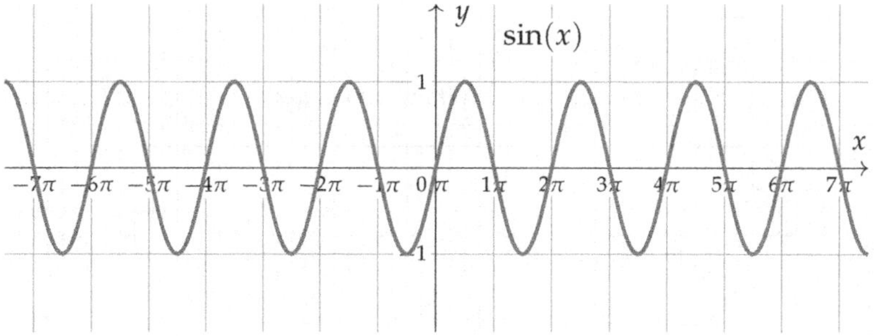

Figure 5.16: The graph of $\sin(x)$ from $x = 0$ to $x = 2\pi$ repeats periodically everywhere else on the number line.

Properties

- Domain: $\mathbb{R}$. The function $f(x) = \sin(x)$ is defined for all input values.
- Image: $\{y \in \mathbb{R} \mid -1 \leqslant y \leqslant 1\}$. The outputs of the sine function are always between -1 and 1.
- Roots: $\{\ldots, -3\pi, -2\pi, -\pi, 0, \pi, 2\pi, 3\pi, \ldots\}$.
  The function $\sin(x)$ has roots at all multiples of π.
- The function is periodic, with period 2π: $\sin(x) = \sin(x + 2\pi)$.
- The sin function is *odd*: $\sin(x) = -\sin(-x)$
- Relation to cos: $\sin^2 x + \cos^2 x = 1$
- Relation to csc: $\csc(x) = \frac{1}{\sin x}$ (csc is read *cosecant*)
- The inverse function of $\sin(x)$ is denoted as $\sin^{-1}(x)$ or $\arcsin(x)$, not to be confused with $(\sin(x))^{-1} = \frac{1}{\sin(x)} = \csc(x)$.
- The number $\sin(\theta)$ is the length-ratio of the vertical side and the hypotenuse in a right-angle triangle with angle θ at the base.

Links

[See the Wikipedia page for nice illustrations]
`http://en.wikipedia.org/wiki/Sine`

Cosine

The cosine function is the same as the sine function *shifted* by $\frac{\pi}{2}$ to the left: $\cos(x) = \sin(x + \frac{\pi}{2})$. Thus everything you know about the sine function also applies to the cosine function.

Graph

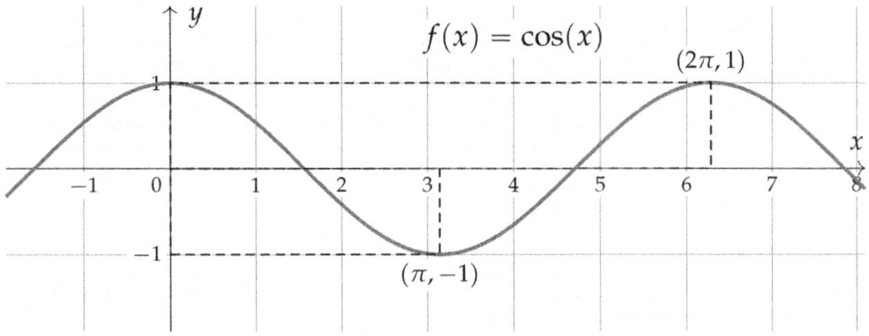

Figure 5.17: The graph of the function $y = \cos(x)$ passes through the following (x, y) coordinates: $(0, 1)$, $(\frac{\pi}{6}, \frac{\sqrt{3}}{2})$, $(\frac{\pi}{4}, \frac{\sqrt{2}}{2})$, $(\frac{\pi}{3}, \frac{1}{2})$, $(\frac{\pi}{2}, 0)$, $(\frac{2\pi}{3}, -\frac{1}{2})$, $(\frac{3\pi}{4}, -\frac{\sqrt{2}}{2})$, $(\frac{5\pi}{6}, -\frac{\sqrt{3}}{2})$, and $(\pi, -1)$.

The cos function starts at $\cos(0) = 1$, then drops down to cross the x-axis at $x = \frac{\pi}{2}$. Cos continues until it reaches its minimum value at $x = \pi$. The function then moves upward, crossing the x-axis again at $x = \frac{3\pi}{2}$, and reaching its maximum value again at $x = 2\pi$.

Properties

- Domain: $\mathbb{R}$
- Image: $\{y \in \mathbb{R} \mid -1 \leqslant y \leqslant 1\}$
- Roots: $\{\ldots, -\frac{3\pi}{2}, -\frac{\pi}{2}, \frac{\pi}{2}, \frac{3\pi}{2}, \frac{5\pi}{2}, \ldots\}$
- Relation to sin: $\sin^2 x + \cos^2 x = 1$
- Relation to sec: $\sec(x) = \frac{1}{\cos x}$ (sec is read *secant*)
- The inverse function of $\cos(x)$ is denoted $\cos^{-1}(x)$ or $\arccos(x)$.
- The cos function is *even*: $\cos(x) = \cos(-x)$
- The number $\cos(\theta)$ is the length-ratio of the horizontal side and the hypotenuse in a right-angle triangle with angle θ at the base

Tangent

The tangent function is the ratio of the sine and cosine functions:

$$f(x) = \tan(x) = \frac{\sin(x)}{\cos(x)}.$$

Graph

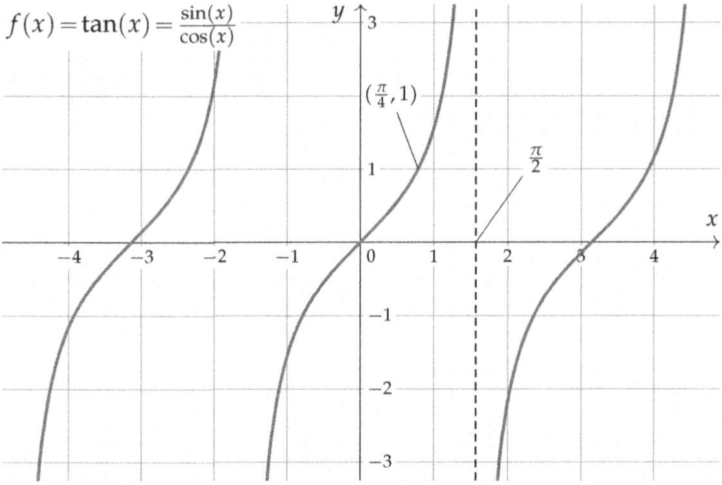

Figure 5.18: The graph of the function $f(x) = \tan(x)$.

Properties

- Domain: $\{x \in \mathbb{R} \mid x \neq \frac{(2n+1)\pi}{2} \text{ for any } n \in \mathbb{Z}\}$
- Image: $\mathbb{R}$
- The function tan is periodic with period π.
- The tan function "blows up" at values of x where $\cos x = 0$. These are called *asymptotes* of the function and their locations are $x = \ldots, \frac{-3\pi}{2}, \frac{-\pi}{2}, \frac{\pi}{2}, \frac{3\pi}{2}, \ldots$.
- Value at $x = 0$: $\tan(0) = \frac{0}{1} = 0$, because $\sin(0) = 0$.
- Value at $x = \frac{\pi}{4}$: $\tan\left(\frac{\pi}{4}\right) = \frac{\sin(\frac{\pi}{4})}{\cos(\frac{\pi}{4})} = \frac{\frac{\sqrt{2}}{2}}{\frac{\sqrt{2}}{2}} = 1$.
- The number $\tan(\theta)$ is the length-ratio of the vertical and the horizontal sides in a right-angle triangle with angle θ.
- The inverse function of $\tan(x)$ is denoted $\tan^{-1}(x)$ or $\arctan(x)$.
- The inverse tangent function is used to compute the angle at the base in a right-angle triangle with horizontal side length ℓ_h and vertical side length ℓ_v: $\theta = \tan^{-1}\left(\frac{\ell_v}{\ell_h}\right)$.

Exponential

The exponential function base $e = 2.7182818\ldots$ is denoted

$$f(x) = e^x = \exp(x).$$

Graph

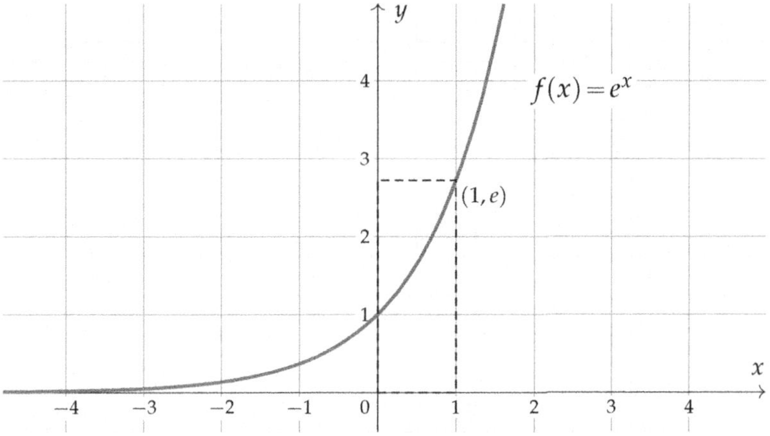

Figure 5.19: The graph of the exponential function $f(x) = e^x$ passes through the following points: $(-2, \frac{1}{e^2})$, $(-1, \frac{1}{e})$, $(0,1)$, $(1,e)$, $(2,e^2)$, $(3,e^3)$, $(4,e^4)$, etc.

Properties

- Domain: $\mathbb{R}$
- Image: $\{y \in \mathbb{R} \mid y > 0\}$
- $f(a)f(b) = f(a+b)$ since $e^a e^b = e^{a+b}$

A more general exponential function would be $f(x) = Ae^{\gamma x}$, where A is the initial value, and γ (the Greek letter *gamma*) is the *rate* of the exponential. For $\gamma > 0$, the function $f(x)$ is increasing, as in Figure 5.19. For $\gamma < 0$, the function is decreasing and tends to zero for large values of x. The case $\gamma = 0$ is special since $e^0 = 1$, so $f(x)$ is a constant of $f(x) = A1^x = A$.

Links

[The exponential function 2^x evaluated]
`http://www.youtube.com/watch?v=e4MSN6IImpI`

Natural logarithm

The natural logarithm function is denoted

$$f(x) = \ln(x) = \log_e(x).$$

The function $\ln(x)$ is the inverse function of the exponential e^x.

Graph

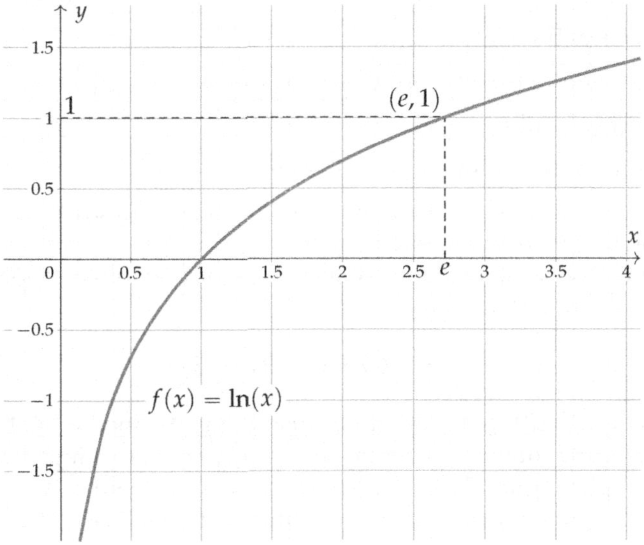

Figure 5.20: The graph of the function $\ln(x)$ passes through the following coordinates: $(\frac{1}{e^2}, -2)$, $(\frac{1}{e}, -1)$, $(1, 0)$, $(e, 1)$, $(e^2, 2)$, $(e^3, 3)$, $(e^4, 4)$, etc.

Properties

- Domain: $\{x \in \mathbb{R} \mid x > 0\}$
- Image: $\mathbb{R}$

Exercises

E5.1 Find the domain, the image, and the roots of $f(x) = 2\cos(x)$.

E5.2 What are the degrees of the following polynomials? Are they even, odd, or neither?

a) $p(x) = x^2 - 5x^4 + 1$ **b)** $q(x) = x - x^3 + x^5 - x^7$

E5.3 Solve for x in the following polynomial equations.

a) $3x + x^2 = x - 15 + 2x^2$ **b)** $3x^2 - 4x - 4 + x^3 = x^3 + 2x + 2$

5.3 Function transformations

Often, we're asked to adjust the shape of a function by scaling it or moving it, so that it passes through certain points. For example, if we wanted to make a function g with the same shape as the absolute value function $f(x) = |x|$, but moved up by three units so that $g(0) = 3$, we would use the function $g(x) = |x| + 3$.

In this section, we'll discuss the four basic transformations you can perform on *any* function f to obtain a transformed function g:

- Vertical translation: $g(x) = f(x) + k$
- Horizontal translation: $g(x) = f(x - h)$
- Vertical scaling: $g(x) = Af(x)$
- Horizontal scaling: $g(x) = f(ax)$

By applying these transformations, we can *move* and *stretch* a generic function to give it any desired shape.

The next couple of pages illustrate all of the above transformations on the function

$$f(x) = 6.75(x^3 - 2x^2 + x).$$

We'll work with this function because it has distinctive features in both the horizontal and vertical directions. By observing this function's graph (Figure 5.21), we see its x-intercepts are at $x = 0$ and $x = 1$. We can confirm this mathematically by factoring the expression:

$$f(x) = 6.75x(x^2 - 2x + 1) = 6.75x(x - 1)^2.$$

The function $f(x)$ also has a local maximum at $x = \frac{1}{3}$, and the value of the function at that maximum is $f(\frac{1}{3}) = 1$.

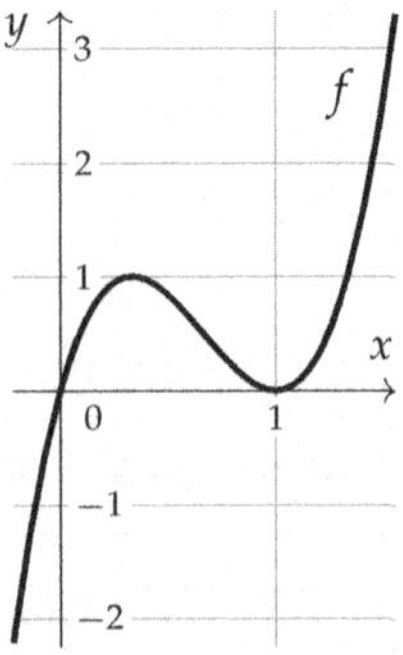

Figure 5.21: Graph of the function $f(x) = 6.75(x^3 - 2x^2 + x)$.

Vertical translations

To move a function $f(x)$ *up* by k units, add k to the function:

$$g(x) = f(x) + k.$$

The function $g(x)$ will have exactly the same shape as $f(x)$, but it will be *translated* (the mathematical term for moved) upward by k units.

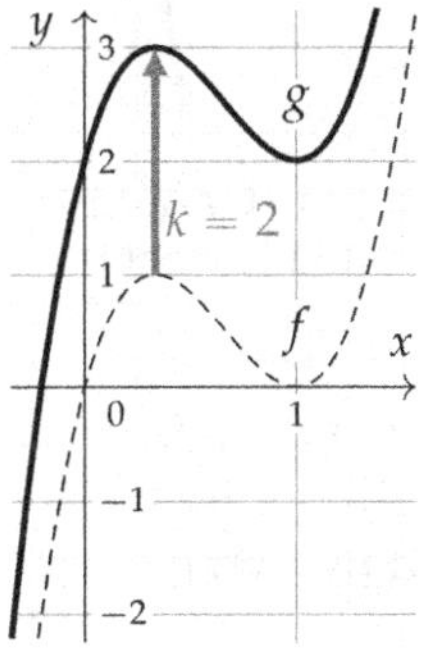

Figure 5.22: The graph of the function $g(x) = f(x) + 2$ has the same shape as the graph of $f(x)$ translated upward by two units.

Recall the function $f(x) = 6.75(x^3 - 2x^2 + x)$. To move the function up by $k = 2$ units, we can write

$$g(x) = f(x) + 2 = 6.75(x^3 - 2x^2 + x) + 2,$$

and the graph of $g(x)$ will be as it is shown in Figure 5.22. Recall the original function $f(x)$ crosses the x-axis at $x = 0$. The transformed function $g(x)$ has the property $g(0) = 2$. The maximum at $x = \frac{1}{3}$ has similarly shifted in value from $f(\frac{1}{3}) = 1$ to $g(\frac{1}{3}) = 3$.

Horizontal translation

We can move a function f to the right by h units by *subtracting* h from x and using $(x - h)$ as the function's input argument:

$$g(x) = f(x - h).$$

The point $(0, f(0))$ on $f(x)$ now corresponds to the point $(h, g(h))$ on $g(x)$.

Figure 5.23 shows the function $f(x) = 6.75(x^3 - 2x^2 + x)$, as well as the function $g(x)$, which is shifted to the right by $h = 2$ units:

$$g(x) = f(x - 2) = 6.75\left[(x-2)^3 - 2(x-2)^2 + (x-2)\right].$$

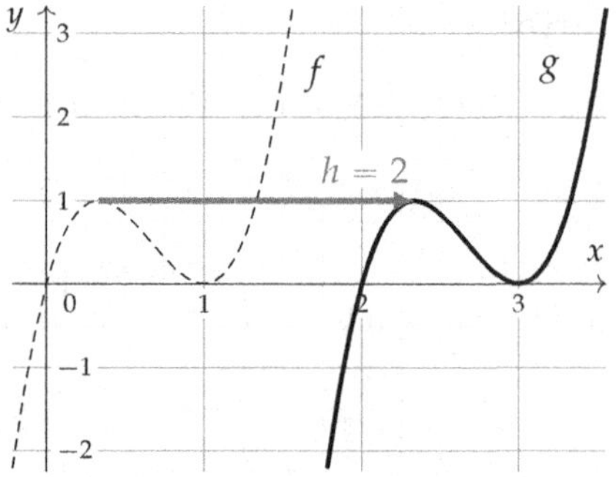

Figure 5.23: The graph of the function $g(x) = f(x-2)$ has the same shape as the graph of $f(x)$ translated to the right by two units.

The original function f gives us $f(0) = 0$ and $f(1) = 0$, so the new function $g(x)$ must give $g(2) = 0$ and $g(3) = 0$. The maximum at $x = \frac{1}{3}$ has similarly shifted by two units to the right, $g(2 + \frac{1}{3}) = 1$.

Vertical scaling

To stretch or compress the shape of a function vertically, we can multiply it by some constant A and obtain

$$g(x) = Af(x).$$

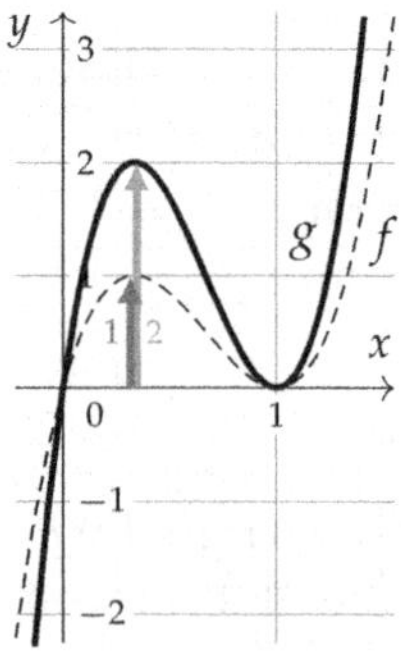

Figure 5.24: The graph of the function $g(x) = 2f(x)$ looks like $f(x)$ vertically stretched by a factor of two.

If $|A| > 1$, the function will be stretched. If $|A| < 1$, the function will be compressed. If A is negative, the function will flip upside down, which is a *reflection* through the x-axis.

There is an important difference between vertical translation and vertical scaling. Translation moves all points of the function by the same amount, whereas scaling moves each point proportionally to that point's distance from the x-axis.

The function $f(x) = 6.75(x^3 - 2x^2 + x)$, when stretched vertically by a factor of $A = 2$, becomes the function

$$g(x) = 2f(x) = 13.5(x^3 - 2x^2 + x).$$

The x-intercepts $f(0) = 0$ and $f(1) = 0$ do not move, and remain at $g(0) = 0$ and $g(1) = 0$. All values of $f(x)$ have been stretched upward by a factor of 2, as we can verify using the point $f(1.5) = 2.5$, which has become $g(1.5) = 5$. The maximum at $x = \frac{1}{3}$ has doubled in value to become $g(\frac{1}{3}) = 2$.

Horizontal scaling

To stretch or compress a function horizontally, we can multiply the input value by some constant a to obtain:

$$g(x) = f(ax).$$

If $|a| > 1$, the function will be compressed. If $|a| < 1$, the function will be stretched. Note that the behaviour here is the opposite of vertical scaling. If a is a negative number, the function will also flip horizontally, which is a reflection through the y-axis.

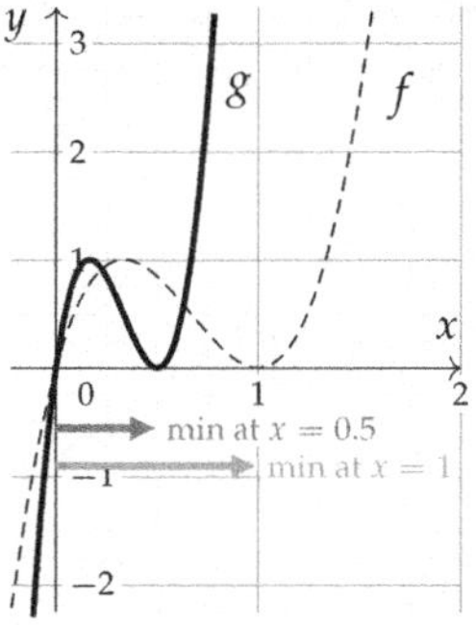

Figure 5.25: The graph of the function $g(x) = f(2x)$ looks like $f(x)$ horizontally compressed by a factor of two.

Figure 5.25 shows the function $g(x)$, which is $f(x)$ compressed horizontally by a factor of $a = 2$:

$$g(x) = f(2x) = 6.75\Big[(2x)^3 - 2(2x)^2 + (2x)\Big].$$

The x-intercept $f(0) = 0$ does not move since it is on the y-axis. The x-intercept $f(1) = 0$ does move, however, and we have $g(0.5) = 0$. The maximum at $x = \frac{1}{3}$ moves to $g(\frac{1}{6}) = 1$. All points of $f(x)$ are compressed toward the y-axis by a factor of 2.

General quadratic function

Any quadratic function can be written in the form:

$$f(x) = a(x-h)^2 + k,$$

where x is the input, and a, h, and k are parameters. This is called the *vertex form* of the quadratic function, and the coordinate pair (h,k) is called the *vertex* of the parabola. This equation can be obtained by starting from the basic quadratic function x^2 (see Figure 5.11) and applying three transformations: a horizontal translation by h units, a vertical scaling by a, and finally a vertical translation by k units.

Parameters

- a: the slope multiplier
  - ▷ The larger the absolute value of a, the steeper the slope.
  - ▷ If $a < 0$ (negative), the function opens downward.
- h: the horizontal displacement of the function. Note that subtracting a number inside the bracket $(\ )^2$ (positive h) makes the function go to the right.
- k: the vertical displacement of the function

The graph in Figure 5.26 illustrates a quadratic function with parameters $a = 1$, $h = 1$ (one unit shifted to the right), and $k = -2$ (two units shifted down).

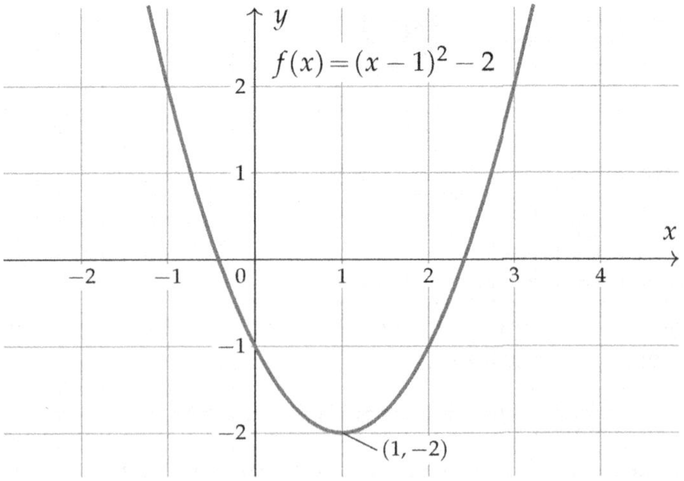

Figure 5.26: The graph of the function $f(x) = (x-1)^2 - 2$ is the same as the function $f(x) = x^2$, but shifted one unit to the right and two units down.

We can also write a quadratic function as a second-degree polynomial $f(x) = ax^2 + bx + c$. This is called the *standard form* of the quadratic function. Given a quadratic expression in standard form $ax^2 + bx + c$, we can find its equivalent expression in vertex form $a(x-h)^2 + k$ using the complete-the-square trick we learned in Section 2.1.

If a quadratic function crosses the x-axis, it can be written in *factored form*:

$$f(x) = a(x - x_1)(x - x_2),$$

where x_1 and x_2 are the two roots of the quadratic. Given a quadratic function $f(x) = ax^2 + bx + c$, we can find its roots using the quadratic formula: $x_1 = \frac{-b+\sqrt{b^2-4ac}}{2a}$ and $x_2 = \frac{-b-\sqrt{b^2-4ac}}{2a}$ (see Section 2.2).

General sine function

Introducing all possible parameters into the sine function gives us:

$$f(x) = A\sin\left(\tfrac{2\pi}{\lambda}x - \phi\right),$$

where A, λ, and ϕ are the function's parameters.

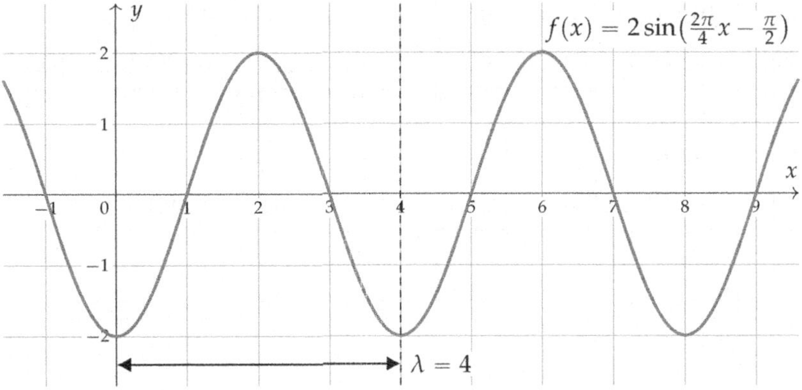

Figure 5.27: The graph of the function $f(x) = 2\sin\left(\frac{2\pi}{4}x - \frac{\pi}{2}\right)$, which has amplitude $A = 2$, wavelength $\lambda = 4$, and phase shift $\phi = \frac{\pi}{2}$.

Parameters

- A: the amplitude describes the distance above and below the x-axis that the function reaches as it oscillates.
- λ: the *wavelength* of the function:

$$\lambda = \{\text{ the horizontal distance from one peak to the next }\}.$$

- ϕ: is a phase shift, analogous to the horizontal shift h, which we have seen. This number dictates where the oscillation starts. The default sine function has zero phase shift ($\phi = 0$), so it passes through the origin with an increasing slope.

The "bare" sine function $f(x) = \sin(x)$ has wavelength 2π and produces outputs that oscillate between -1 and $+1$. When we multiply the bare function by the constant A, the oscillations will range between $-A$ and A. When the input x is scaled by the factor $\frac{2\pi}{\lambda}$, the wavelength of the function becomes λ.

Exercises

E5.4 Given the functions $f(x) = x + 5$, $g(x) = x - 6$, $h(x) = 7x$, and $q(x) = x^2$, find the formulas for the following composite functions:

a) $q \circ f$ **b)** $f \circ q$ **c)** $q \circ g$ **d)** $q \circ h$

In each case, describe how the graph of the composite function is related to the graph of $q(x)$.
Hint: Recall, "$\circ$" denotes function composition: $(f \circ g)(x) = f(g(x))$.

E5.5 Find the amplitude A, the wavelength λ, and the phase shift ϕ for the function $f(x) = 5\sin(62.83x - \frac{\pi}{8})$.

E5.6 Choose the coefficients a, b, and c for the quadratic function $f(x) = ax^2 + bx + c$ so that it passes through the points $(0,5)$, $(1,4)$, and $(2,5)$.
Hint: Find the equation $f(x) = A(x - h)^2 + k$ first.

E5.7 Find the values α and β that will make the function $g(x) = 2\sqrt{x - \alpha} + \beta$ pass through the points $(3,-2)$, $(4,0)$, and $(7,2)$.

Chapter 6

Geometry

Geometry is the mathematical study of shapes and proportions. This branch of mathematics encompasses all things that can be drawn and analyzed using geometric concepts like similarity, proportionality, and orthogonality. Geometry studies the common, abstract nature of all geometrical shapes. For example, instead of studying and describing the properties of any particular triangle, we'll study the geometrical properties common to all triangles.

We won't just look at shapes; we'll develop a language for describing them quantitatively in terms of lengths and angles. We can find patterns in different geometric shapes and describe proportions and lengths using equations.

Here's the chapter breakdown. In Section 6.1, we'll discuss properties of several geometric shapes: triangles, circles, cylinders, spheres, cones, and pyramids. Sections 6.2 and 6.3 discuss the topic of *trigonometry*, which is the study of the proportions of the sides of right-angle triangles. We'll learn all about the trigonometric functions $\sin\theta$ and $\cos\theta$, which play an important role in many areas of mathematics. In Section 6.4 we'll learn more about circles and *polar coordinates*. Sections 6.5, 6.6, and 6.7 will discuss the *ellipse*, the *parabola*, and the *hyperbola*, which are three other interesting geometric shapes that occur in nature.

6.1 Geometry formulas

The word "geometry" comes from the Greek roots *geo*, which means "earth," and *metron*, which means "measurement." This name is linked to one of the early applications of geometry, which was to measure the total amount of land contained within a certain boundary region. Over the years, the study of geometry evolved to be more abstract. Instead of developing formulas for calculating the area of specific regions of land, mathematicians developed general area formulas that apply to *all* regions that have a particular shape.

In this section we'll present formulas for calculating the perimeters, areas, and volumes for various shapes (also called "figures") commonly encountered in the real world. For two-dimensional figures, the main quantities of interest are the figures' areas and the figures' perimeters (the length of the walk around the figure). For three-dimensional figures, the quantities of interest are the surface area (how much paint it would take to cover all sides of the figure), and volume (how much water it would take to fill a container of this shape). The formulas presented are by no means an exhaustive list of everything there is to know about geometry, but they represent a core set of facts that you want to add to your toolbox.

Triangles

The area of a triangle is equal to $\frac{1}{2}$ times the length of its base times its height:

$$A = \tfrac{1}{2}ah_a.$$

Note that h_a is the height of the triangle *relative to* the side a.

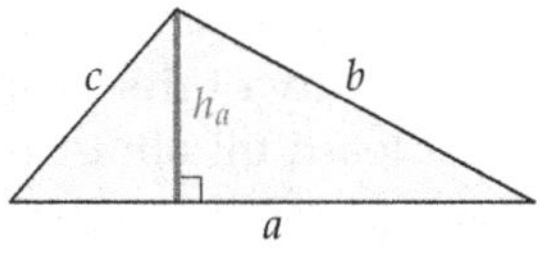

Figure 6.1: A triangle with side lengths a, b, and c. The height of the triangle with respect to the side a is denoted h_a.

The perimeter of a triangle is given by the sum of its side lengths:

$$P = a + b + c.$$

Interior angles of a triangle rule The sum of the inner angles in any triangle is equal to $180°$. Consider a triangle with internal angles α, β and γ as shown in Figure 6.2. We may not know the values of the individual angles α, β, and γ, but we know their sum is $\alpha + \beta + \gamma = 180°$.

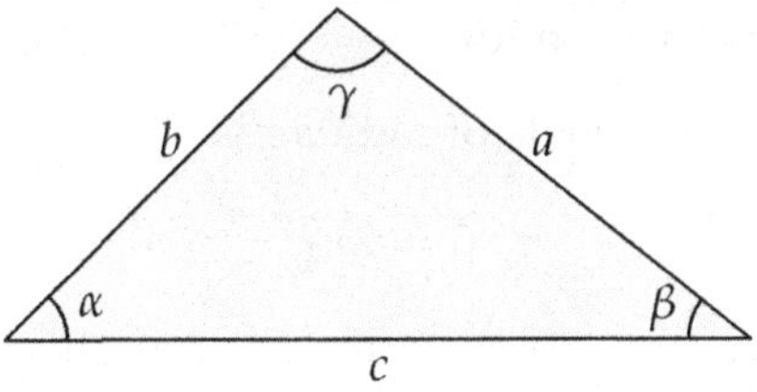

Figure 6.2: A triangle with inner angles α, β, and γ and sides a, b, and c.

Sine rule The sine rule states the following equation is true:

$$\frac{a}{\sin(\alpha)} = \frac{b}{\sin(\beta)} = \frac{c}{\sin(\gamma)},$$

where α is the angle opposite to side a, β is the angle opposite to side b, and γ is the angle opposite to side c, as shown in Figure 6.2.

Cosine rule The cosine rules states the following equations are true:

$$a^2 = b^2 + c^2 - 2bc\cos(\alpha),$$
$$b^2 = a^2 + c^2 - 2ac\cos(\beta),$$
$$c^2 = a^2 + b^2 - 2ab\cos(\gamma).$$

These equations are useful when you know two sides of a triangle and the angle between them, and you want to find the third side.

Circle

The circle is a beautiful shape. If we take the centre of the circle at the origin $(0,0)$, the circle of radius r corresponds to the equation

$$x^2 + y^2 = r^2.$$

This formula describes the set of points (x,y) with a distance from the centre equal to r.

Area

The area enclosed by a circle of radius r is given by $A = \pi r^2$. A circle of radius $r = 1$ has area π.

Circumference and arc length

The circumference of a circle of radius r is

$$C = 2\pi r.$$

A circle of radius $r = 1$ has circumference 2π. This is the total length you can measure by following the curve all the way around to trace the outline of the entire circle. For example, the circumference of a circle of radius 3 m is $C = 2\pi(3) = 18.85$ m. This is how far you'll need to walk to complete a full turn around a circle of radius $r = 3$ m.

What is the length of a part of the circle? Say you have a piece of the circle, called an *arc*, and that piece corresponds to the angle $\theta = 57°$ as shown in Figure 6.3. What is the arc's length ℓ? If the circle's total length $C = 2\pi r$ represents a full 360° turn around the circle, then the arc length ℓ for a portion of the circle corresponding to the angle θ is

$$\ell = 2\pi r \frac{\theta}{360}.$$

The arc length ℓ depends on r, the angle θ, and a factor of $\frac{2\pi}{360}$.

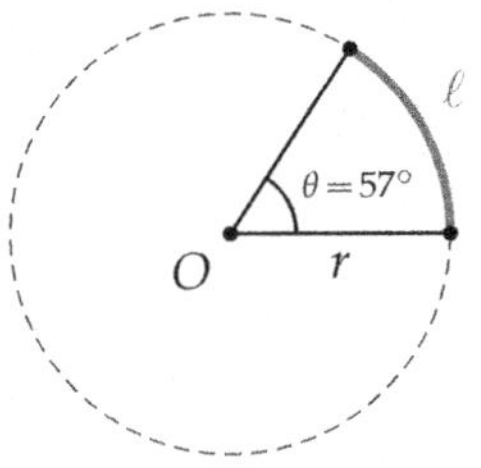

Figure 6.3: The arc length ℓ equals $\frac{57}{360}$ of the circle's circumference $2\pi r$.

Radians

While scientists and engineers commonly use degrees as a measurement unit for angles, mathematicians prefer to measure angles in *radians*, denoted rad.

Measuring an angle in radians is equivalent to measuring the arc length ℓ on a circle with radius $r = 1$, as illustrated in Figure 6.4. The conversion ratio between degrees and radians is

$$2\pi \text{ rad} = 360°.$$

When the angle θ is measured in radians, the arc length is given by:

$$\ell = r\theta.$$

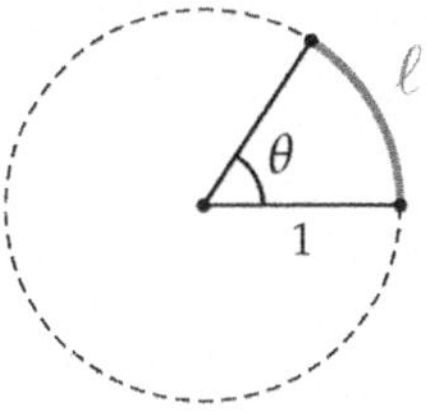

Figure 6.4: The angle θ measured in radians corresponds to the arc length ℓ on a circle with radius 1. The full circle corresponds to the angle 2π rad.

To find the arc length ℓ, we simply multiply the circle radius r times the angle θ measured in radians.

Note the arc-length formula with θ measured in radians is simpler than the arc-length formula with θ measured in degrees, since we don't need the conversion factor of $360°$.

The geometry of circles is so important that we dedicated a whole section (Section 6.4) to pursuing this topic in more detail. For now, let's continue discussing some other important geometric shapes.

Sphere

A sphere of radius r is described by the equation $x^2 + y^2 + z^2 = r^2$. The surface area of the sphere is $A = 4\pi r^2$, and its volume is given by $V = \frac{4}{3}\pi r^3$.

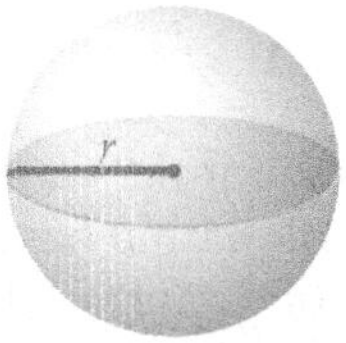

Figure 6.5: A sphere of radius r has surface area $4\pi r^2$ and volume $\frac{4}{3}\pi r^3$.

Cylinder

The surface area of a cylinder consists of the top and bottom circular surfaces, plus the area of the side of the cylinder:

$$A = 2\left(\pi r^2\right) + (2\pi r)h.$$

The volume of a cylinder is the product of the area of the cylinder's base times its height:

$$V = \left(\pi r^2\right) h.$$

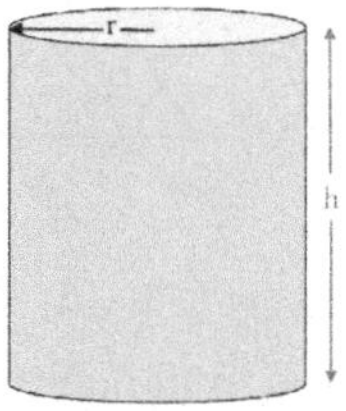

Figure 6.6: A cylinder with radius r and height h has volume $\pi r^2 h$.

Example You open the hood of your car and see 2.0 L written on top of the engine. The 2.0 L refers to the combined volume of the four pistons, which are cylindrical in shape. The owner's manual tells you the radius of each piston is 43.75 mm, and the height of each piston is 83.1 mm. Verify the total engine volume is 1998789 $\text{mm}^3 \approx 2$ L.

Cones and pyramids

The volume of a square pyramid with side length a and height h is given by the formula $V = \frac{1}{3}a^2h$. The volume of a cone of radius r and height h is given by the formula $V = \frac{1}{3}\pi r^2 h$. Note the factor $\frac{1}{3}$ appears in both formulas. These two formulas are particular cases of the general volume formula that applies to all pyramids:

$$V = \frac{1}{3}Ah,$$

where A is the area of the pyramid's base and h is its height. This formula applies for pyramids with a base that is a triangle (triangular pyramids), a square (square pyramids), a rectangle (rectangular pyramids), a circle (cones), or any other shape.

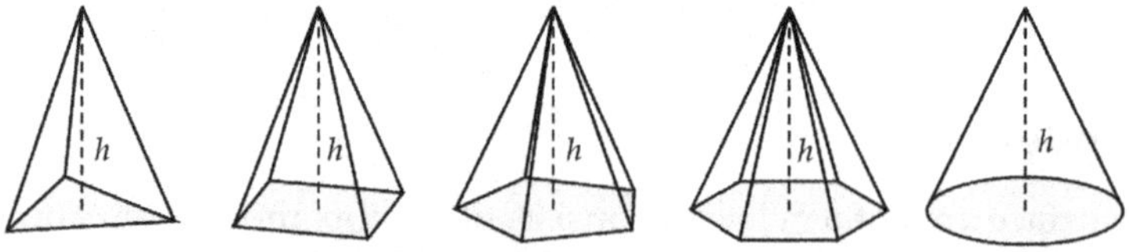

Figure 6.7: The volumes of pyramids and cones are described by the formula $V = \frac{1}{3}Ah$, where A is the area of the base and h is the height.

Exercises

E6.1 Find the length of side x in the triangle below.

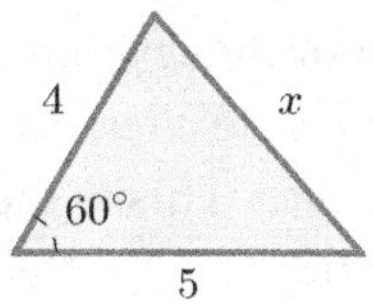

Hint: Use the cosine rule.

E6.2 Find the volume and the surface area of a sphere with radius 2.

E6.3 On a rainy day, Laura brings her bike indoors, and the wet bicycle tires leave a track of water on the floor. What is the length of the water track left by the bike's rear tire (diameter 73 cm) if the wheel makes five full turns along the floor?

6.2 Trigonometry

If one of the angles in a triangle is equal to 90°, we call this triangle a *right-angle triangle.* In this section we'll discuss right-angle triangles in great detail and get to know their properties. We'll learn some fancy new terms like *hypotenuse, opposite,* and *adjacent,* which are used to refer to the different sides of a triangle. We'll also use the functions *sine, cosine,* and *tangent* to compute the *ratios of lengths* in right triangles.

Understanding triangles and their associated trigonometric functions is of fundamental importance: you'll need this knowledge for your future understanding of mathematical concepts like vectors and complex numbers.

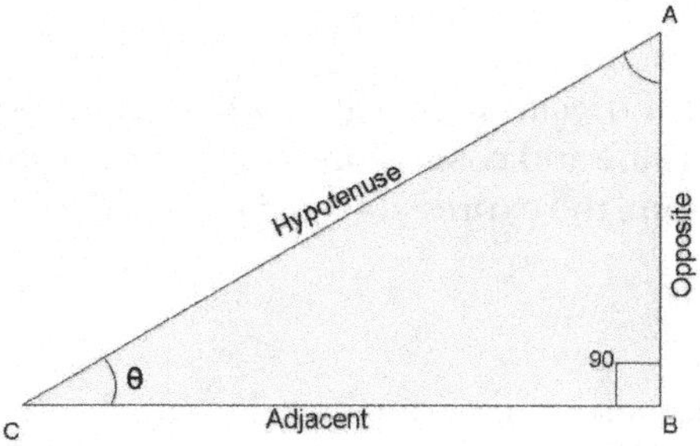

Figure 6.8: A right-angle triangle. The angle at the base is denoted θ and the names of the sides of the triangle are indicated.

Concepts

- A, B, C: the three *vertices* of the triangle
- θ: the angle at the vertex C. Angles can be measured in degrees or radians.

- opp $= AB$: the length of the *opposite* side to θ
- adj $= BC$: the length of side *adjacent* to θ
- hyp $= AC$: the *hypotenuse*. This is the triangle's longest side.
- h: the "height" of the triangle (in this case $h = \text{opp} = AB$)
- $\sin\theta = \frac{\text{opp}}{\text{hyp}}$: the *sine* of theta is the ratio of the length of the opposite side and the length of the hypotenuse
- $\cos\theta = \frac{\text{adj}}{\text{hyp}}$: the *cosine* of theta is the ratio of the adjacent length and the hypotenuse length
- $\tan\theta = \frac{\sin\theta}{\cos\theta} = \frac{\text{opp}}{\text{adj}}$: the *tangent* is the ratio of the opposite length divided by the adjacent length

Pythagoras' theorem

In a right-angle triangle, the length of the hypotenuse squared is equal to the sum of the squares of the lengths of the other sides:

$$\text{adj}^2 + \text{opp}^2 = \text{hyp}^2.$$

If we divide both sides of the above equation by hyp^2, we obtain

$$\frac{\text{adj}^2}{\text{hyp}^2} + \frac{\text{opp}^2}{\text{hyp}^2} = 1.$$

Since $\frac{\text{adj}}{\text{hyp}} = \cos\theta$ and $\frac{\text{opp}}{\text{hyp}} = \sin\theta$, this equation can be rewritten as

$$\cos^2\theta + \sin^2\theta = 1.$$

This is a powerful *trigonometric identity* that describes an important relation between sine and cosine functions. In case you've never seen this notation before, the expression $\cos^2\theta$ is used to denote $(\cos(\theta))^2$.

Sin and cos

Meet the trigonometric functions, or trigs for short. These are your new friends. Don't be shy now, say hello to them.

"Hello."

"Hi."

"Soooooo, you are like functions right?"

"Yep," sin and cos reply in chorus.

"Okay, so what do you do?"

"Who me?" asks cos. "Well I tell the ratio...hmm...Wait, are you asking what I do as a *function* or specifically what *I* do?"

"Both I guess?"

"Well, as a function, I take angles as inputs and I give ratios as answers. More specifically, I tell you how 'wide' a triangle with that angle will be," says cos all in one breath.

"What do you mean wide?" you ask.

"Oh yeah, I forgot to say, the triangle must have a hypotenuse of length 1. What happens is there is a point P that moves around on a circle of radius 1, and we *imagine* a triangle formed by the point P, the origin, and the point on the x-axis located directly below the point P."

"I am not sure I get it," you confess.

"Let me try explaining," says sin. "Look at Figure 6.9 and you'll see a circle. This is the unit circle because it has a radius of 1. You see it, yes?"

"Yes."

"Now imagine a point P that moves along the circle of radius 1, starting from the point $P(0) = (1,0)$. The x and y coordinates of the point $P(\theta) = (P_x(\theta),\, P_y(\theta))$ as a function of θ are

$$P(\theta) = (P_x(\theta),\, P_y(\theta)) = (\cos\theta,\, \sin\theta).$$

So, *either* you can think of us in the context of triangles, or in the context of the unit circle."

"Cool. I kind of get it. Thanks so much," you say, but in reality you are weirded out. Talking functions? "Well guys. It was nice to meet you, but I have to get going, to finish the rest of the book."

"See you later," says cos.

"Peace out," says sin.

The unit circle

The *unit circle* is a circle of radius one centred at the origin. The unit circle consists of all points (x, y) that satisfy the equation $x^2 + y^2 = 1$. A point P on the unit circle has coordinates $(P_x, P_y) = (\cos\theta, \sin\theta)$, where θ is the angle P makes with the x-axis.

Figure 6.10 shows the graph of the function $f(\theta) = \sin\theta$. The values $\sin\theta$ for the angles 0, $\frac{\pi}{6}$ (30°), $\frac{\pi}{3}$ (60°), and $\frac{\pi}{2}$ (90°) are marked. There are three values to remember: $\sin\theta = 0$ when $\theta = 0$, $\sin\theta = \frac{1}{2}$ when $\theta = \frac{\pi}{6}$ (30°), and $\sin\theta = 1$ when $\theta = \frac{\pi}{2}$ (90°). See Figure 5.14 (page 80) for a graph of $\sin\theta$ that shows a complete cycle around the circle. Also see Figure 5.17 (page 82) for the graph of $\cos\theta$.

Instead of trying to memorize the values of the functions $\cos\theta$ and $\sin\theta$ separately, it's easier to remember them as a combined "package" $(\cos\theta, \sin\theta)$, which describes the x- and y-coordinates of the point P for the angle θ. Figure 6.11 shows the values of $\cos\theta$ and $\sin\theta$ for the angles 0, $\frac{\pi}{6}$ (30°), $\frac{\pi}{4}$ (45°), $\frac{\pi}{3}$ (60°), and $\frac{\pi}{2}$ (90°). These

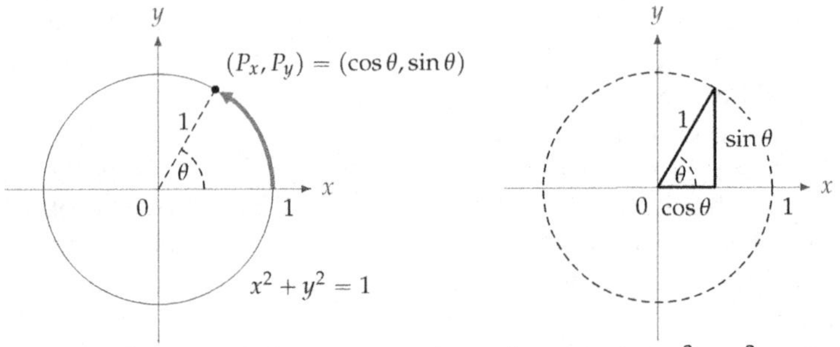

Figure 6.9: The unit circle corresponds to the equation $x^2 + y^2 = 1$. The coordinates of the point P on the unit circle are $P_x = \cos\theta$ and $P_y = \sin\theta$.

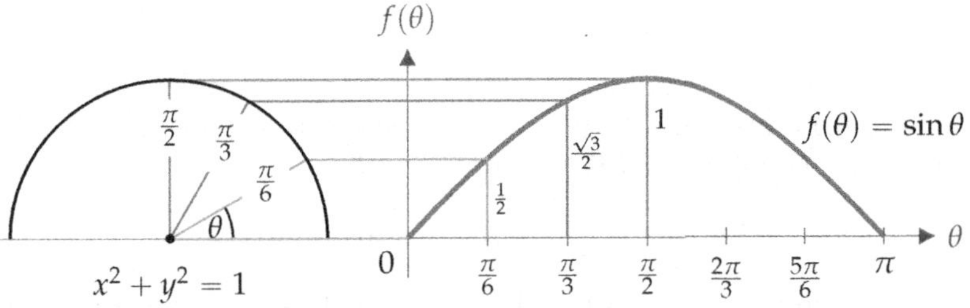

Figure 6.10: The function $f(\theta) = \sin\theta$ describes the vertical position of a point P that travels along the unit circle. The graph shows the values of the function $f(\theta) = \sin\theta$ for angles between $\theta = 0$ and $\theta = \pi$.

are the most common angles that often show up on homework and exam questions. For each angle, the x-coordinate (the first number in the bracket) is $\cos\theta$, and the y-coordinate (the second number in the bracket) is $\sin\theta$.

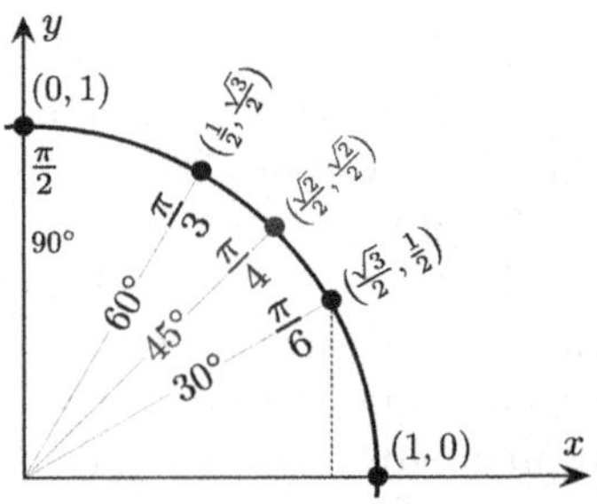

Figure 6.11: The combined $(\cos\theta, \sin\theta)$ coordinates for the points on the unit circle at the most common angles: 0, $\frac{\pi}{6}$ (30°), $\frac{\pi}{4}$ (45°), $\frac{\pi}{3}$ (60°), and $\frac{\pi}{2}$ (90°).

Note the values of $\cos\theta$ and $\sin\theta$ for the angles shown in Figure 6.11 are all combinations of the fractions $\frac{1}{2}$, $\frac{\sqrt{2}}{2}$, and $\frac{\sqrt{3}}{2}$. The square roots appear as a consequence of the trigonometric identity $\cos^2\theta + \sin^2\theta = 1$. This identity tells us that the sum of the squared coordinates of each point on the unit circle is equal to one. Let's look at what this equation tells us for the angle $\theta = \frac{\pi}{6}$ (30°). Remember that $\sin(30°) = \frac{1}{2}$ (the length of the dashed line in Figure 6.11). We can plug this value into the equation $\cos^2(30°) + \sin^2(30°) = 1$ to find the value: $\cos(30°) = \sqrt{1 - \sin^2(30°)} = \sqrt{1 - \frac{1}{4}} = \sqrt{\frac{3}{4}} = \frac{\sqrt{3}}{2}$.

The coordinates $\left(\frac{\sqrt{2}}{2}, \frac{\sqrt{2}}{2}\right)$ for the angle $\theta = \frac{\pi}{4}$ (45°) are obtained from a similar calculation. We know the values of $\sin\theta$ and $\cos\theta$ must be equal for that angle, so we're looking for the number a that satisfies the equation $a^2 + a^2 = 1$, which is $a = \frac{1}{\sqrt{2}} = \frac{\sqrt{2}}{2}$. The values of $\cos(60°)$ and $\sin(60°)$ can be obtained from a symmetry argument. Measuring 60° from the x-axis is the same as measuring 30° from the y-axis, so $\cos(60°) = \sin(30°) = \frac{1}{2}$ and $\sin(60°) = \cos(30°) = \frac{\sqrt{3}}{2}$.

We can extend the calculations described above for all other angles that are multiples of $\frac{\pi}{6}$ (30°) and $\frac{\pi}{4}$ (45°) to obtain the $\cos\theta$ and $\sin\theta$ values for the whole unit circle, as shown in Figure 6.12.

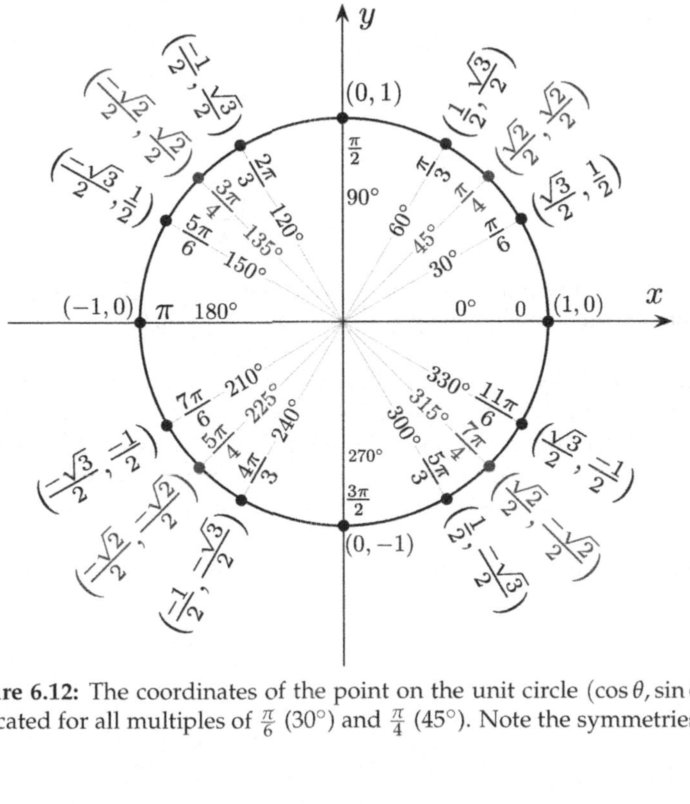

Figure 6.12: The coordinates of the point on the unit circle $(\cos\theta, \sin\theta)$ are indicated for all multiples of $\frac{\pi}{6}$ (30°) and $\frac{\pi}{4}$ (45°). Note the symmetries.

Don't be intimidated by all the information shown in Figure 6.12! You're not expected to memorize all these values. The primary reason for including this figure is so you can appreciate the symmetries of the sine and cosine values that we find as we go around the circle. The values of $\sin\theta$ and $\cos\theta$ for all angles are the same as the values for the angles between 0° and 90°, but one or more of their coordinates has a negative sign. For example, 150° is just like 30°, except its x-coordinate is negative since the point lies to the left of the y-axis. Another use for Figure 6.12 is to convert between angles measured in degrees and radians, since both units for angles are indicated.

Non-unit circles

Consider a point $Q(\theta)$ at an angle of θ on a circle with radius $r \neq 1$. How can we find the x- and y-coordinates of the point $Q(\theta)$?

We saw that the coefficients $\cos\theta$ and $\sin\theta$ correspond to the x- and y-coordinates of a point on the *unit* circle ($r = 1$). To obtain the coordinates for a point on a circle of radius r, we must *scale* the coordinates by a factor of r:

$$Q(\theta) = (Q_x(\theta), Q_y(\theta)) = (r\cos\theta, r\sin\theta).$$

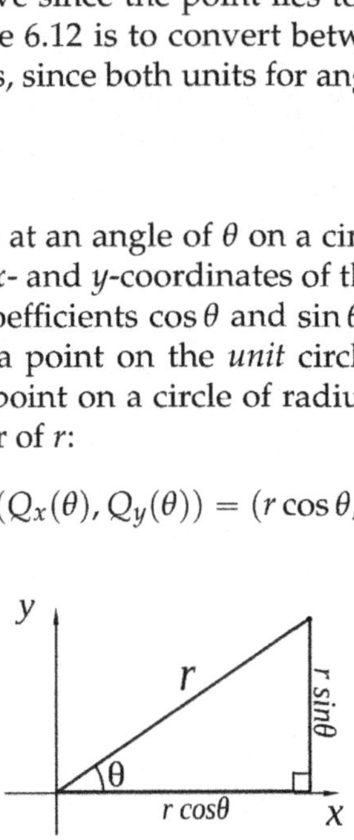

Figure 6.13: The x- and y-coordinates of a point at the angle θ and distance of r from the origin are given by $x = r\cos\theta$ and $y = r\sin\theta$.

The take-away message is that you can use the functions $\cos\theta$ and $\sin\theta$ to find the "horizontal" and "vertical" components of any length r. From this point on in the book, we'll always talk about the length of the *adjacent* side as $x = r\cos\theta$, and the length of the *opposite* side as $y = r\sin\theta$. It is extremely important you get comfortable with this notation.

The reasoning behind the above calculations is as follows:

$$\cos\theta = \frac{\text{adj}}{\text{hyp}} = \frac{x}{r} \quad\Rightarrow\quad x = r\cos\theta,$$

and

$$\sin\theta = \frac{\text{opp}}{\text{hyp}} = \frac{y}{r} \quad\Rightarrow\quad y = r\sin\theta.$$

Calculators

Watch out for the units of angle measures when using calculators and computers. Make sure you know what kind of angle units the functions sin, cos, and tan expect as inputs, and what kind of outputs the functions $\sin^{-1}$, $\cos^{-1}$, and $\tan^{-1}$ return.

For example, let's see what we should type into the calculator to compute the sine of 30 degrees. If the calculator is set to degrees, we simply type: `3`, `0`, `sin`, `=`, and obtain the answer 0.5.

If the calculator is set to radians, we have two options:

1. Change the `mode` of the calculator so it works in degrees.
2. Convert $30°$ to radians

$$30° \times \frac{2\pi \text{ rad}}{360°} = \frac{\pi}{6} \text{ rad},$$

   and type: `π`, `/`, `6`, `sin`, `=` on the calculator.

Try computing $\cos(60°)$, $\cos\left(\frac{\pi}{3} \text{ rad}\right)$, and $\cos^{-1}\left(\frac{1}{2}\right)$ using your calculator to make sure you know how it works.

Exercises

E6.4 Given a circle with radius $r = 5$, find the x- and y-coordinates of the point at $\theta = 45°$. What is the circumference of the circle?

E6.5 Convert the following angles from degrees to radians.

a) $30°$ **b)** $45°$ **c)** $60°$ **d)** $270°$

Links

[Unit-circle walkthrough and tricks by patrickJMT on YouTube]
`http://bit.ly/1mQg9Cj` and `http://bit.ly/1hvA702`

6.3 Trigonometric identities

There are a number of important relationships between the values of the functions sin and cos. Here are three of these relationships, known as *trigonometric identities*. There about a dozen other identities that are less important, but you should memorize these three.

The three identities to remember are:

1. Unit hypotenuse

$$\sin^2\theta + \cos^2\theta = 1.$$

The unit hypotenuse identity is true by the Pythagoras theorem and the definitions of sin and cos. The sum of the squares of the sides of a triangle is equal to the square of the hypotenuse.

2. Sine angle sum

$$\sin(a+b) = \sin(a)\cos(b) + \sin(b)\cos(a).$$

The mnemonic for this identity is "sico + sico."

3. Cosine angle sum

$$\cos(a+b) = \cos(a)\cos(b) - \sin(a)\sin(b).$$

The mnemonic for this identity is "coco − sisi." The negative sign is there because it's not good to be a sissy.

Derived formulas

If you remember the above three formulas, you can derive pretty much all the other trigonometric identities.

Double angle formulas

Starting from the sico + sico identity and setting $a = b = x$, we can derive the following identity:

$$\sin(2x) = 2\sin(x)\cos(x).$$

Starting from the coco-sisi identity, we obtain

$$\begin{aligned} \cos(2x) &= \cos^2(x) - \sin^2(x) \\ &= 2\cos^2(x) - 1 = 2\left(1 - \sin^2(x)\right) - 1 = 1 - 2\sin^2(x). \end{aligned}$$

The formulas for expressing $\sin(2x)$ and $\cos(2x)$ in terms of $\sin(x)$ and $\cos(x)$ are called *double angle formulas*.

If we rewrite the double-angle formula for $\cos(2x)$ to isolate the $\sin^2$ or the $\cos^2$ term, we obtain the *power-reduction formulas*:

$$\cos^2(x) = \frac{1}{2}(1 + \cos(2x)), \qquad \sin^2(x) = \frac{1}{2}(1 - \cos(2x)).$$

Self-similarity

Sin and cos are periodic functions with period 2π. Adding a multiple of 2π to the function's input does not change the function:

$$\sin(x+2\pi)=\sin(x), \qquad \cos(x+2\pi)=\cos(x).$$

This follows because adding a multiple of 2π brings us back to the same point on the unit circle.

Furthermore, sin and cos have symmetries with respect to zero,

$$\sin(-x)=-\sin(x), \qquad \cos(-x)=\cos(x),$$

within each π half-cycle,

$$\sin(\pi-x)=\sin(x), \qquad \cos(\pi-x)=-\cos(x),$$

and within each full 2π cycle,

$$\sin(2\pi-x)=-\sin(x), \qquad \cos(2\pi-x)=\cos(x).$$

Take the time to revisit Figure 5.14 (page 80), Figure 5.17 (page 82), and Figure 6.12 (page 103) to visually confirm that all the equations shown above are true. Knowing the points where the functions take on the same values (symmetries) or take on opposite values (anti-symmetries) is very useful in calculations.

Sin is cos, cos is sin

It shouldn't be surprising if I tell you that sin and cos are actually $\frac{\pi}{2}$-shifted versions of each other:

$$\cos(x)=\sin\left(x+\tfrac{\pi}{2}\right), \qquad \sin(x)=\cos\left(x-\tfrac{\pi}{2}\right).$$

Formulas for sums and products

Here are some formulas for transforming sums into products:

$$\sin(a)+\sin(b)=2\sin\left(\tfrac{1}{2}(a+b)\right)\cos\left(\tfrac{1}{2}(a-b)\right),$$

$$\sin(a)-\sin(b)=2\sin\left(\tfrac{1}{2}(a-b)\right)\cos\left(\tfrac{1}{2}(a+b)\right),$$

$$\cos(a)+\cos(b)=2\cos\left(\tfrac{1}{2}(a+b)\right)\cos\left(\tfrac{1}{2}(a-b)\right),$$

$$\cos(a)-\cos(b)=-2\sin\left(\tfrac{1}{2}(a+b)\right)\sin\left(\tfrac{1}{2}(a-b)\right).$$

And here are some formulas for transforming products into sums:

$$\sin(a)\cos(b)=\tfrac{1}{2}\Big(\sin(a+b)+\sin(a-b)\Big),$$

$$\sin(a)\sin(b)=\tfrac{1}{2}\Big(\cos(a-b)-\cos(a+b)\Big),$$

$$\cos(a)\cos(b)=\tfrac{1}{2}\Big(\cos(a-b)+\cos(a+b)\Big).$$

Discussion

The above formulas will come in handy when you need to find some unknown in an equation, or when you are trying to simplify

Exercises

E6.6 Given $a = \pi$ and $b = \frac{\pi}{2}$, find

a) $\sin(a+b)$ **b)** $\cos(2a)$ **c)** $\cos(a+b)$

E6.7 Simplify the following expressions and compute their value without using a calculator.

a) $\cos(x) + \cos(\pi - x)$ **b)** $2\sin^2(x) + \cos(2x)$

c) $\sin(\frac{5\pi}{4})\sin(-\frac{\pi}{4})$ **d)** $2\cos(\frac{5\pi}{4})\cos(-\frac{\pi}{4})\cos(\pi)$

6.4 Circles and polar coordinates

In this section, we'll review what we know about circles and define the *polar coordinate system*, a specialized coordinate system for describing circles and other circular shapes.

Formulas

A *circle* is a set of points located at a constant distance from a centre point. A circle with radius r centred at the origin is described by the equation

$$x^2 + y^2 = r^2.$$

All points (x, y) that satisfy this equation are part of the circle.

More generally, the circle's centre can be located at any point (h, k) in the plane, as illustrated in Figure 6.14.

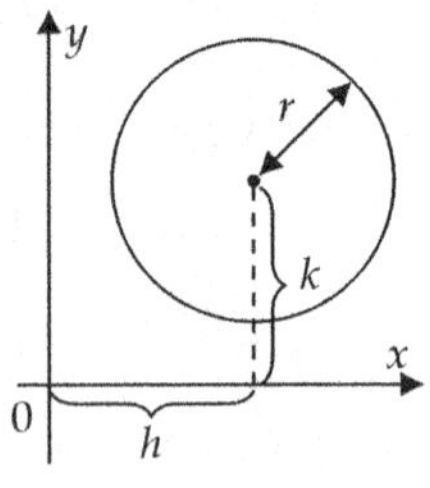

Figure 6.14: A circle of radius r centred at the point (h, k) is described by the equation $(x-h)^2 + (y-k)^2 = r^2$.

Describing circles using functions

The circle equation $x^2 + y^2 = r^2$ is a *relation* between the variables x and y. If we want to describe the circle using a function $y = f(x)$, we can solve for y in the equation $x^2 + y^2 = r^2$ to obtain

$$y = f_t(x) = \sqrt{r^2 - x^2}, \qquad -r \leqslant x \leqslant r,$$

and

$$y = f_b(x) = -\sqrt{r^2 - x^2}, \quad -r \leqslant x \leqslant r.$$

Describing a circle requires two functions, f_t and f_b, because there are two values of y that satisfy the equation $x^2 + y^2 = r^2$ for each value of x. The function f_t describes the top half of the circle, while the function f_b describes the bottom half.

You might be wondering why a simple geometric shape like a circle requires such complicated-looking formulas like $f_t(x)$ and $f_b(x)$ to describe it. Surely there's a better way to describe circles that doesn't involve quadratic expressions and square roots? There is! If instead of using the Cartesian coordinates (x, y) we use the polar coordinates $r\angle\theta$, then the equation of a circle becomes very simple. We'll learn about that next.

The polar coordinate system

Figure 6.15 shows the *polar coordinate system*, which consists of concentric circles at different distances from the origin (also called the *pole*), and radial lines extending from the origin in all directions. We can specify the location of any point in the plane using the *polar coordinates* $r\angle\theta$, where r measures the point's distance from the origin, and θ describes the angle measured in the counterclockwise direction starting from the r-axis. For example, the point $Q = 2\angle 60°$ is located at the distance of $r = 2$ units form the origin, in the direction $\theta = 60°$.

Compare the polar coordinate system shown in Figure 6.15 with the Cartesian coordinate system in Figure 4.3. In the Cartesian coordinate system, we interpret the coordinate pair (x, y) as the instructions "Walk a distance of x units in the direction of the x-axis, and a distance of y units in the direction of the y-axis." In a polar coordinate system, we interpret the coordinates $r\angle\theta$ as the instructions "Turn toward the direction θ and walk a distance of r units in that direction." Both types of coordinates give instructions for getting to a particular point in the plane, with Cartesian coordinates giving the instructions in the form of two distances, while polar coordinates give the instructions in the form of a distance and a direction.

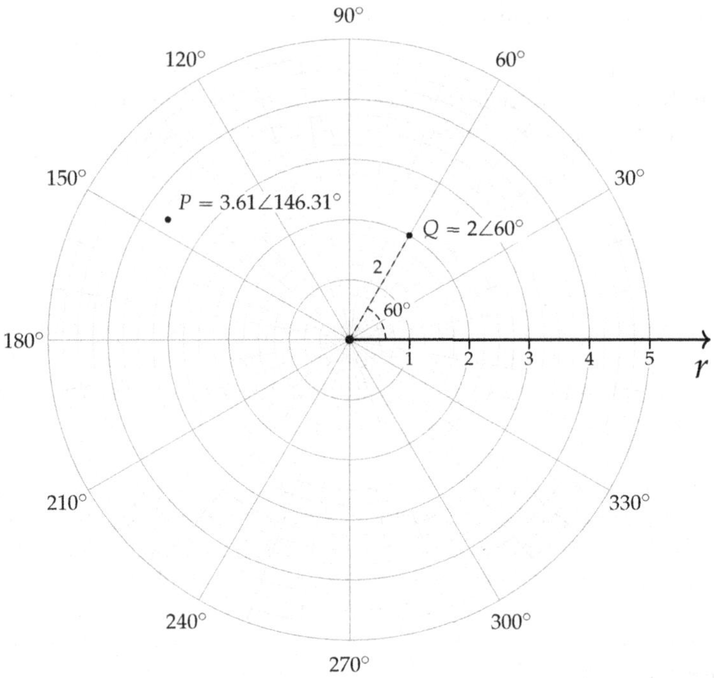

Figure 6.15: We can use the polar coordinate system to describe points in the two-dimensional plane. The polar coordinates $r\angle\theta$ describe the point located at the distance r from the origin in the direction θ.

A Cartesian coordinate pair (x, y) is made of x and y coordinates, while a polar coordinate pair $r\angle\theta$ is made of r and θ coordinates. In this book, we use the angle symbol $\angle$ (read "at an angle of") to separate the polar coordinates r and θ, in order to emphasize the difference between Cartesian and polar coordinates. However, some other books use the notation (r, θ) for polar coordinates, so you have to watch out—the coordinate pair $(20, 30)$ could be either a (x, y) coordinate pair or a (r, θ) coordinate pair, depending on the context.

Note the polar coordinates that describe a given point are not unique, meaning the same point can be described in multiple ways. The point $Q = 2\angle 60°$ is equally described by the polar coordinates $2\angle - 300°$, since a clockwise turn of $300°$ is the same as a counterclockwise turn of $60°$. We can also describe the same point Q using the polar coordinates $-2\angle 240°$ and $-2\angle - 120°$, which tell us to turn in the direction opposite to $60°$ and measure a negative distance $r = -2$. While all of these polar coordinates for Q are equivalent, the preferred way to specify polar coordinates is with positive r values and angles $|\theta| \leqslant 180°$.

Converting between Cartesian and polar coordinates

Figure 6.16 shows a point whose location is described both in terms of Cartesian coordinates (x,y) and polar coordinates $r\angle\theta$. The triangle formed by the coordinates $(0,0)$, $(x,0)$, and (x,y) is a right-angle triangle. This means we can apply our knowledge of the trigonometric functions sin, cos, and tan to obtain formulas for converting between Cartesian coordinates (x,y) and polar coordinates $r\angle\theta$.

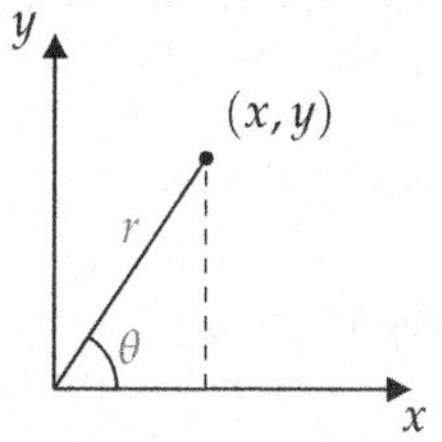

Figure 6.16: Polar coordinates $r\angle\theta$ can describe any point (x,y).

To convert from polar coordinates $r\angle\theta$ to (x,y) coordinates, we use the definitions of the trigonometric functions $\cos\theta = \frac{\text{adj}}{\text{hyp}} = \frac{x}{r}$ and $\sin\theta = \frac{\text{opp}}{\text{hyp}} = \frac{y}{r}$ to obtain the formulas:

$$x = r\cos\theta \qquad \text{and} \qquad y = r\sin\theta.$$

For example, the Cartesian coordinates of the point $Q = 2\angle 60°$ are given by $Q = (x,y) = (2\cos 60°, 2\sin 60°) = (1, \sqrt{3})$.

To convert from (x,y) coordinates to $r\angle\theta$ coordinates, we can use the circle equation $x^2 + y^2 = r^2$ and the definition of the tangent function $\tan\theta = \frac{\text{opp}}{\text{adj}} = \frac{y}{x}$, then solve for r and θ to obtain the formulas:

$$r = \sqrt{x^2 + y^2} \quad \text{and} \quad \theta = \begin{cases} \tan^{-1}\left(\frac{y}{x}\right) & \text{if } x > 0, \\ 180° + \tan^{-1}\left(\frac{y}{x}\right) & \text{if } x < 0, \\ 90° & \text{if } x = 0 \text{ and } y > 0, \\ -90° & \text{if } x = 0 \text{ and } y < 0. \end{cases}$$

Finding the angle θ is a little tricky. We must use a different formula for computing θ depending on where the point is located, and there are four different cases to consider. The basic idea is to use the inverse tangent function $\tan^{-1}$, which is also called arctan, or `atan` on computer systems. By convention, the function $\tan^{-1}$ returns values between $-90°$ ($-\frac{\pi}{2}$ rad) and $90°$ ($\frac{\pi}{2}$ rad), which correspond to points with positive x-coordinates. If the x-coordinate of the point is negative, we must add $180°$ (π rad) to the output of the inverse-tangent calculation to obtain the correct angle. When $x = 0$ we can't compute the fraction $\frac{y}{x}$ because we cannot divide by zero, so we must handle the cases with $x = 0$ separately as described in the above equation.

If you have access to a computer algebra system, the easiest way to calculate the angle θ for the point (x, y) is to use the two-input inverse tangent function `atan2(y,x)`. The function `atan2` is the best way to compute the angle since it handles all four cases of converting Cartesian coordinates to polar coordinates automatically and always gives the correct angle. You can try some calculations with `atan2` using the computer algebra system at `https://live.sympy.org`.

For example, consider the point P with Cartesian coordinates $(-3, 2)$ shown in Figure 4.3 (page 55). To find the polar coordinates of this point we first calculate the distance from the centre, $r = \sqrt{(-3)^2 + 2^2} = \sqrt{13} \approx 3.61$. To find the angle θ we note that the x-coordinate of P is negative, so the angle θ we're looking for is given by the formula $\theta = 180° + \tan^{-1}(\frac{2}{-3}) = 146.31°$. The angle of the point $P = (-3, 2)$ can also be obtained from `atan2(2,-3)`. The polar coordinates of the point P are $3.61\angle 146.31°$ (see Figure 6.15).

Equations in polar coordinates

Equations in polar coordinates serve to describe relations between the variables r and θ. For example, the equation of a circle with radius 2 in polar coordinates is simply $r = 2$. If we substitute $r = \sqrt{x^2 + y^2}$ and square both sides of the equation, we obtain the equation $x^2 + y^2 = 2^2$ that we saw in the beginning of this section.

We can use the substitutions $x = r\cos\theta$ and $y = r\sin\theta$ to convert equations from Cartesian coordinates x and y to polar coordinates r and θ. Consider the equation $2x - y = 3$, which describes the line shown in Figure 5.10 on page 72. We can rewrite this equation in polar coordinates as $2r\cos\theta - r\sin\theta = 3$, which is a relation between the polar coordinates r and θ.

As you can tell from these examples, polar coordinates are very convenient when dealing with circles, and less so when working with lines. Indeed, describing a circle in polar coordinates is as simple as $r = 2$, while in Cartesian coordinates we had to use the complicated-looking functions f_t and f_b (see page 109). The situation is the opposite for lines: the equation of a line in Cartesian coordinates is simple, $2x - y = 3$, while in polar coordinates the same line is described by a tangled mess involving sin and cos functions.

Functions in polar coordinates

A function in polar coordinates is denoted $r(\theta)$ and describes how the distance r varies as a function of the angle θ.

For example, a circle with radius 2 is described by the function $r(\theta) = 2$ in polar coordinates, as illustrated in Figure 6.17 (a). The circle is described by a constant function in polar coordinates, since the points in all directions have the same distance from the centre.

As another example, we can transform the equation of the line $2x - y = 3$ to polar coordinates to obtain $2r\cos\theta - r\sin\theta = 3$, then isolate r to obtain the function

$$r(\theta) = \frac{3}{2\cos\theta - \sin\theta},$$

which describes the distance from the origin for different angles θ. For example, when $\theta = 0$, we find $r(0) = \frac{3}{2\cos 0 - \sin 0} = 1.5$, so we can plot the polar coordinates $1.5\angle 0^\circ$ on the function's graph.

The polar coordinates graph of the function $r(\theta)$ corresponds to all points with polar coordinates $r(\theta)\angle\theta$, for all possible values of θ. This is analogous to how we obtain the graph of the function $f(x)$ in Cartesian coordinates by plotting the points $(x, f(x))$, for all possible values of the input variable x. Figure 6.17 shows the graphs of the two functions discussed here.

If you ever need to graph a function $r(\theta)$ by hand, you can compute the value of the function for several angles like $\theta = -90^\circ$, $\theta = 0^\circ, \theta = 90^\circ$, then plot these points in the polar coordinate system. For example, to graph the function $r(\theta) = \frac{3}{2\cos\theta - \sin\theta}$, we can compute $r(-90^\circ) = \frac{3}{2\cos(-90^\circ) - \sin(-90^\circ)} = 3$, which gives us the point $3\angle -90^\circ$ on the graph. We can similarly compute $r(0^\circ) = 1.5$ and $r(30^\circ) = 2.43$, which gives us the points $1.5\angle 0^\circ$ and $2.43\angle 30^\circ$.

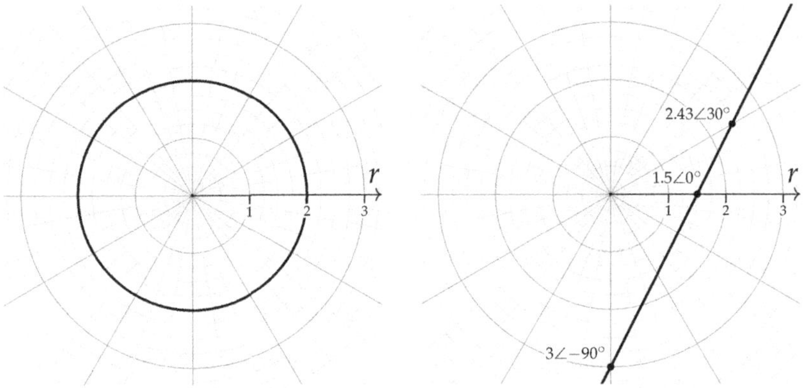

(a) $r(\theta) = 2$ (b) $r(\theta) = \frac{3}{2\cos\theta - \sin\theta}$

Figure 6.17: The graphs of functions in polar coordinates are obtained by computing the distance $r(\theta)$ in all directions θ varying from 0° to 360°.

Figure 6.18 shows the polar coordinates graphs of three other interesting functions. Look at the points $r\angle\theta$ indicated in each graph and check that they satisfy the corresponding function $r(\theta)$.

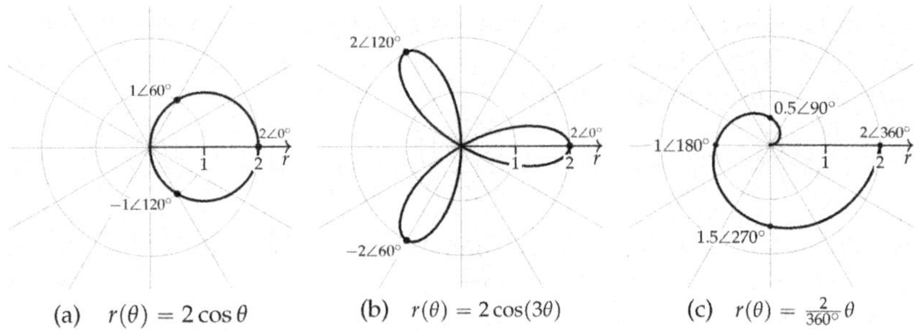

Figure 6.18: The graphs of three functions in polar coordinates: (a) a circle, (b) a three-petalled rose, and (c) an Archimedean spiral.

Discussion

The polar coordinate system is an alternative way of describing points in space using polar coordinates $r\angle\theta$ instead of the usual Cartesian coordinates (x, y). See the concept map in Figure 6.19. Your knowledge and experience with the trigonometric functions sin, cos, and tan is what allows you to convert between Cartesian and polar coordinates.

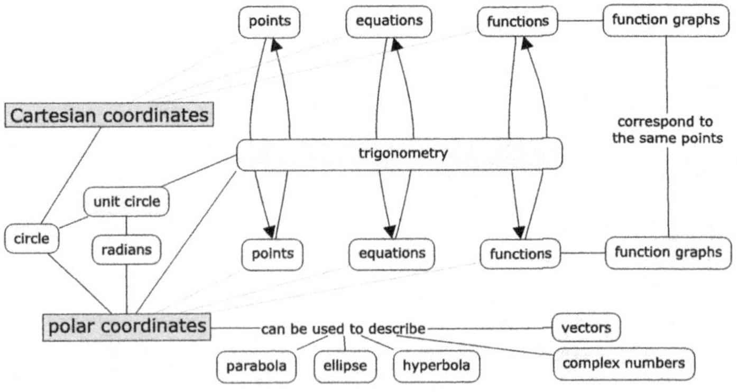

Figure 6.19: Cartesian coordinates (x, y) and polar coordinates $r\angle\theta$ are two equivalent systems for representing points, equations, and functions.

The formulas for converting between Cartesian coordinates (x, y) and polar coordinates $r\angle\theta$ covered in this section are important, and you should consider them "required material." I expect you to become totally fluent with these formulas now, because we'll need them later in the book when we learn about vectors (Section 7.2) and complex numbers (Section 7.5).

In contrast, the three next sections are not "required material." We'll now switch gears to "entertainment mode" and learn about three bonus geometry topics: ellipses, parabolas, and hyperbolas. I want you to know about these shapes, but I don't expect you to be fluent with all the definitions and equations. You can take it easy for the next three sections because none of the material will be "on the exam." You deserve a break after all the polar coordinates formulas!

Exercises

E6.8 Convert the given points from Cartesian to polar coordinates:

a) $(3,1)$ **b)** $(-1,-2)$ **c)** $(0,-6)$

Convert the points from polar to Cartesian coordinates:

d) $10\angle 30°$ **e)** $10\angle -345°$ **f)** $10\angle 120°$

E6.9 Draw the graph of the function $r(\theta) = \frac{2}{\sin\theta}$ in polar coordinates for θ varying from 0 to 180°. What is the equivalent description of this function in Cartesian coordinates?

Links

[Visual introduction to polar coordinates]
`https://www.youtube.com/watch?v=stU63ST6ung`

[Professor Dave explains equations in polar coordinates]
`https://www.youtube.com/watch?v=jwLUapqnwkk`

6.5 Ellipse

The *ellipse* is a fundamental shape that occurs in nature. The orbit of planet Earth around the Sun is an ellipse.

Parameters

Figure 6.20 shows an ellipse with all its parameters annotated:

- F_1, F_2: the two *focal points* of the ellipse
- r_1: the distance from a point on the ellipse to F_1
- r_2: the distance from a point on the ellipse to F_2
- a: the semi-major axis of the ellipse is the half-length of the ellipse along the x-axis. The distance between V_1 and V_2 is $2a$.

- b: the semi-minor axis of the ellipse is the half-width of the ellipse along the y-axis. The distance between V_3 and V_4 is $2b$.
- c: the distance of the focal points from the centre of the ellipse. The distance between F_1 and F_2 is $2c$.
- ε: the *eccentricity* of the ellipse, $\varepsilon = \sqrt{1 - \frac{b^2}{a^2}} = \frac{c}{a}$.

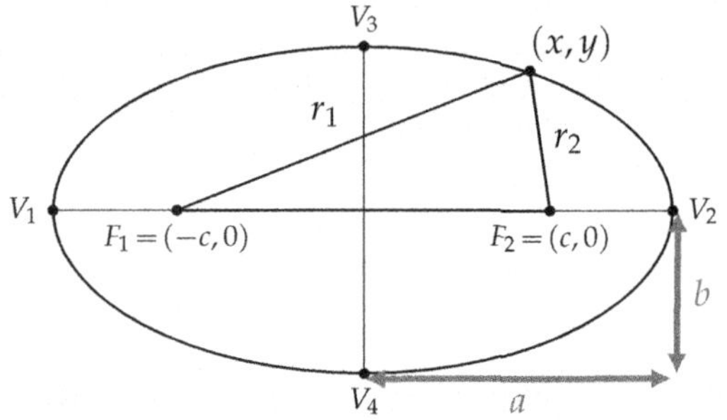

Figure 6.20: An ellipse with semi-major axis a and semi-minor axis b. The locations of the focal points F_1 and F_2 are indicated.

Definition

An ellipse is the curve found by tracing along all the points for which the sum of the distances to the two focal points is a constant:

$$r_1 + r_2 = \text{const.}$$

There's a neat way to draw a perfect ellipse using a piece of string and two tacks. Take a piece of string and tack it to a picnic table at two points, leaving some loose slack in the middle of the string. Now take a pencil, and without touching the table, use the pencil to pull the middle of the string until it is taut. Make a mark at that point. With the two parts of string completely straight, make a mark at every point possible where the two "legs" of string remain taut.

An ellipse is a set of points (x, y) that satisfy the equation

$$\frac{x^2}{a^2} + \frac{y^2}{b^2} = 1.$$

The parameters a and b determine the shape of the ellipse.

The focal points F_1 and F_2 correspond to the locations of the two tacks where the string is held in place. The coordinates of the two focal points are

$$F_1 = (-c, 0) \qquad \text{and} \qquad F_2 = (c, 0),$$

where $c = \sqrt{a^2 - b^2}$ is the *focal distance*.

The *eccentricity* of an ellipse is given by the equation

$$\varepsilon = \sqrt{1 - \frac{b^2}{a^2}} = \frac{c}{a}.$$

The parameter ε (the Greek letter *epsilon*) varies between 0 and 1 and describes how much the shape of the ellipse differs from the shape of a circle. When $\varepsilon = 0$ the ellipse is a circle with radius a, and both focal points are located at the centre. As the eccentricity ε increases, the ellipse becomes more elongated and the focal points spread farther apart.

Polar coordinates

Consider a polar coordination system whose centre is located at the focus F_2. We can describe the ellipse by specifying the function $r_2(\theta)$, which describes the distance from the focus F_2 to the point E on the ellipse as a function of the angle θ (see Figure 6.21). Recall that for functions in polar coordinates, the angle θ is the independent variable that varies from 0 to 2π (360°), and the dependent variable is the distance $r_2(\theta)$.

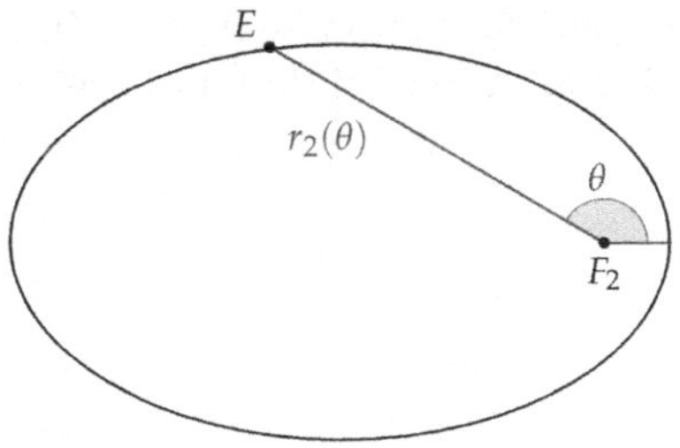

Figure 6.21: The function $r_2(\theta)$ in polar coordinates specifies the distance between the point E on the ellipse and the focal point F_2 for all angles.

The function that describes an ellipse in polar coordinates is

$$r_2(\theta) = \frac{a(1 - \varepsilon^2)}{1 + \varepsilon\cos(\theta)},$$

where the angle θ is measured with respect to the semi-major axis. The distance is smallest when $\theta = 0$ with $r_2(0) = a - c = a(1 - \varepsilon)$ and largest when $\theta = \pi$ with $r_2(\pi) = a + c = a(1 + \varepsilon)$.

Calculating the orbit of the Earth

The motion of the Earth around the Sun is an ellipse with the Sun positioned at the focus F_2. We can therefore use the polar coordinates formula $r_2(\theta)$ to describe the distance of the Earth from the Sun. The eccentricity of Earth's orbit around the Sun is $\varepsilon = 0.01671123$, and the half-length of the major axis is $a = 149\,598\,261$ km. We substitute these values into the general formula for $r_2(\theta)$ and obtain the following equation:

$$r_2(\theta) = \frac{149\,556\,484}{1 + 0.01671123\cos(\theta)} \text{ km.}$$

The point where the Earth is closest to the Sun is called the *perihelion*. It occurs when $\theta = 0$, which happens around the 3rd of January. The moment where the Earth is most distant from the Sun is called the *aphelion* and corresponds to the angle $\theta = \pi$. Earth's *aphelion* happens around the 3rd of July.

Let's use the formula for $r_2(\theta)$ to predict the *perihelion* and *aphelion* distances of Earth's orbit:

$$r_{2,\text{peri}} = r_2(0) = \frac{149556483}{1 + 0.01671123\cos(0)} = 147\,098\,290 \text{ km},$$

$$r_{2,\text{aphe}} = r_2(\pi) = \frac{149556483}{1 + 0.01671123\cos(\pi)} = 152\,098\,232 \text{ km}.$$

Google "perihelion" and "aphelion" to verify that the above predictions are accurate. It's kind of cool that a mathematical formula can describe the motion of our planet, don't you think?

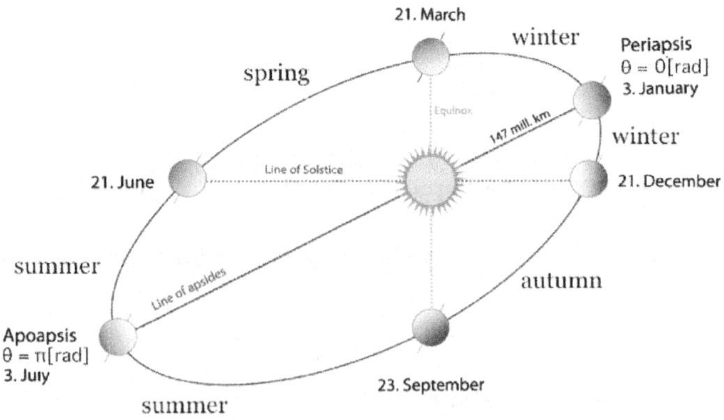

Figure 6.22: The orbit of the Earth around the Sun. Key points of the orbit are labelled. The seasons in the Northern hemisphere are also indicated.

The angle θ of the Earth relative to the Sun can be described as a function of time $\theta(t)$. The exact formula of the function $\theta(t)$ that describes the angle as a function of time is fairly complicated, so we won't go into the details. Let's simply look at the values of $\theta(t)$ with t measured in days shown in Table 6.2. We'll begin on Jan 3rd.

t in days	1	2	$\cdots$	182	$\cdots$	365	365.2422
date	Jan 3	Jan 4	$\cdots$	July 3	$\cdots$	Jan 2	?
$\theta(t)$ in $^\circ$	0		$\cdots$	180	$\cdots$	359.7614	360
$\theta(t)$ in rad	0		$\cdots$	π	$\cdots$	6.2790	2π

Table 6.2: The angular position of the Earth as a function of time. Note the extra amount of "day" that is roughly equal to $\frac{1}{4} = 0.25$. We account for this discrepancy by adding an extra day to the calendar once every four years.

Newton's insight

Contrary to common belief, Newton did not discover his theory of gravitation because an apple fell on his head while sitting under a tree. What actually happened is that he started from Kepler's laws of motion, which describe the exact elliptical orbit of the Earth as a function of time. Newton asked, "What kind of force would cause two bodies to spin around each other in an elliptical orbit?" He determined that the gravitational force between the Sun of mass M and the Earth of mass m must be of the form $F_g = \frac{GMm}{r^2}$.

For now, let's give props to Newton for connecting the dots, and props to Johannes Kepler for studying the orbital periods, and Tycho Brahe for doing all the astronomical measurements. Above all, we owe some props to the ellipse for being such an awesome shape!

Exercises

E6.10 The focal points of the ellipse with equation $\frac{x^2}{a^2} + \frac{y^2}{b^2} = 1$ are $F_1 = (-c, 0)$ and $F_2 = (c, 0)$, as illustrated in Figure 6.20. Use the definition of the ellipse $r_1 + r_2 =$ const. to compute the value of the parameter c in terms of the parameters a and b.

Links

[Interactive graph of an ellipse]
`https://www.desmos.com/calculator/kgmh67lroj`

[Further reading about Earth-Sun geometry]
`http://www.physicalgeography.net/fundamentals/6h.html`

6.6 Parabola

The *parabola* is another important geometric shape. In this section, we'll see how we can describe parabolas using their geometric properties, as well as in terms of algebraic equations.

Parameters

Figure 6.23 shows a parabola with all its parameters annotated:

- f: the *focal length* of the parabola
- $F = (0, f)$: the *focal point* of the parabola
- $\{(x, y) \in \mathbb{R}^2 \mid y = -f\}$: the *directrix* line to the parabola
- r: the distance from point P on the parabola to the focal point F
- ℓ: the closest distance from a point P on the parabola to the parabola's directrix line

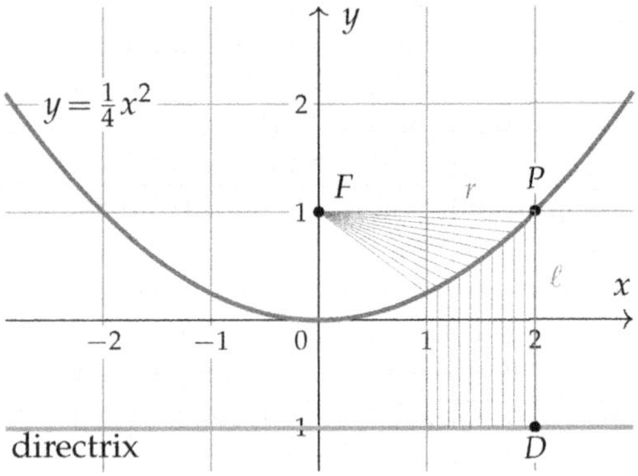

Figure 6.23: The parabola is defined geometrically as the set of points whose distance form the focal point r is equal to their distance from the directrix ℓ. The figure shows the point P on the parabola that has distance $r = 2$ from F, and distance $\ell = 2$ from the point D on the directrix. This parabola can be described algebraically using the equation $y = \frac{1}{4}x^2$.

Geometric definition

The shape of a parabola is determined by a single parameter f, called the *focal length*. For a parabola with focal length f, the focal point is at $F = (0, f)$ and the *directrix* line has the equation $y = -f$. The parabola is defined as the set of points P for which the distance from the focal point and the directrix are equal:

$$r = \ell,$$

where $r = d(P, F)$ is the distance from the point P to the focal point F, and $\ell = d(P, D)$ is the distance from P to the point D on the directrix that is closest to the point P.

Figure 6.23 shows a parabola opening upward with focal length $f = 1$ centred at the origin. The parabola is the set of points that are equidistant from the focal point $F = (0, 1)$ and the directrix line located at $y = -1$.

Algebraic description

The shape of a parabola with focal length f opening upward corresponds to the graph of the quadratic functions $f(x) = \frac{1}{4f}x^2$. This is a special case of the general formula for quadratic functions $f(x) = a(x-h)^2 + k$, which you're already familiar with from Section 5.3 (see page 90). The parabola shown in Figure 6.23 is centred at the origin, so the displacement parameters h and k are both zero. The coefficient a in the general formula is related to the focal length f through the relation $a = \frac{1}{4f}$, so in the case of focal length $f = 1$ the coefficient is $a = \frac{1}{4}$. See Figure 6.23.

The formula $y = \frac{1}{4f}x^2$ is specific to the case of a parabola opening upward, but similar algebraic expressions exist for parabolas opening downward and sideways. The parabola with focal length f opening downward is described by the equation $y = -\frac{1}{4f}x^2$. The parabola opening to the left and to the right are described by relations $x = -\frac{1}{4f}y^2$ and $x = \frac{1}{4f}y^2$. With your knowledge of the displacement parameters h and k used for general quadratic equations (see page 90), you can also obtain algebraic expressions for parabolas that are not centred at the origin.

Polar coordinates

In the previous section we connected the geometric definition of parabolas with quadratic algebraic expressions. When learning math, it's important to note connections of this sort because they are the bridges between different mathematical domains. If one day you have to solve a geometry problem involving parabolas, you could use algebraic equations to describe the parabolas and solve the problem using algebra. If on another day you encounter an algebra problem involving a quadratic equation, you could visualize the quadratic equation as a parabolic shape and solve the problem using geometric reasoning. Being able to travel between math domains like this is a mark of true math fluency.

In the spirit of further bridge-building, I want to show you the equation of a parabola in polar coordinates. We choose a coordinate system centred at the focal point F. The polar-coordinates equation for the parabola with focal length f opening to the left is

$$r(\theta) = \frac{2f}{1 + \cos\theta}.$$

Figure 6.24 shows a particular instance of this formula when the parabola has focal length $f = 1$. Try substituting the values $\theta = 0$ and $\theta = 90°$ ($\frac{\pi}{2}$ radians) in the polar equation to verify that it correctly describes the points on the parabola.

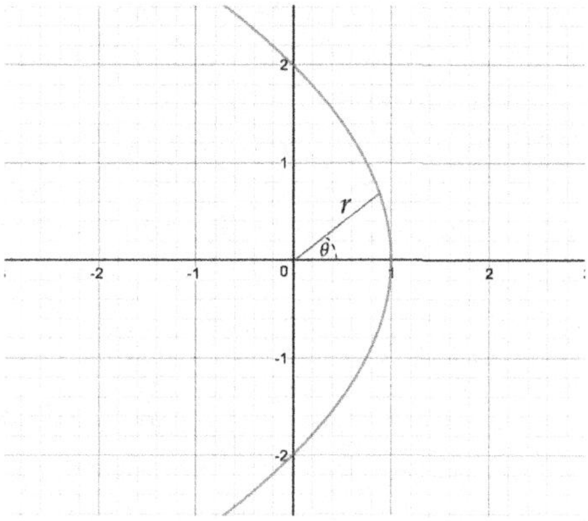

Figure 6.24: The parabola described by $r(\theta) = \frac{2}{1+\cos\theta}$ in polar coordinates.

The key point I want you to take away from this section is that algebraic formulas can be very useful for describing geometric shapes. The parabola illustrated in Figure 6.24 can be described in three equivalent ways: geometrically through its focal length $f = 1$ and directrix line $x = 2$; algebraically as the relation $x = 1 - \frac{1}{4}y^2$ in Cartesian coordinates; or as the function $r(\theta) = \frac{2}{1+\cos\theta}$ in polar coordinates.

Parabola applications

Parabolic shapes are of special importance in optics and communications. Using parabolic lenses, mirrors, and antennas, it's possible to focus the energy emitted from a distant object into a single point. This is due to the *reflective property* of parabolas, which states that all light rays coming from far away are redirected toward the focal point of the parabolic shape. The reflective property makes parabolas useful for many practical communication applications.

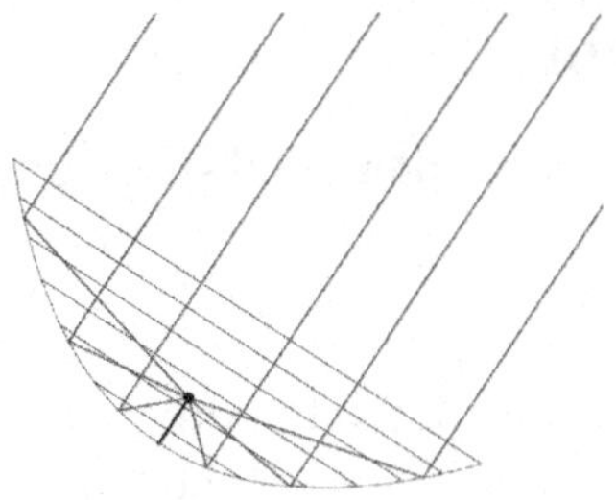

Figure 6.25: The reflective property of parabolas tells us all radio waves coming from infinity are reflected toward the focal point of the parabola.

Figure 6.25 illustrates the setup for a radio communication scenario in which a ground station is trying to detect a signal coming from a satellite in orbit. The satellite is very far away so the signal received on Earth is very weak. A parabolic satellite dish antenna collects the signal from a large surface area and focuses all of it on the focal point of the parabola. A radio receiver placed at the focal point of the parabola receives a much stronger signal, since the focal point is where the power from the whole dish surface is concentrated. This is thanks to the reflective property of the parabolic shape: all radio waves coming from the far-away satellite get reflected toward the focal point of the parabola.

Exercises

E6.11 Consider some arbitrary point $P = (x, y)$ that lies on the parabola with focal length f centred at the origin as illustrated in Figure 6.23. Use the geometric definition of the parabola $r = \ell$ to obtain a relation between the x- and y-coordinates of the point P.

Hint: The distance between points $A = (A_x, A_y)$ and $B = (B_x, B_y)$ is given by $d(A, B) = \sqrt{(A_x - B_x)^2 + (A_y - B_y)^2}$.

Hint: Recall the definitions of $r = d(P, F)$ and $\ell = d(P, D)$.

Links

[Interactive graph of a parabola]
`https://www.desmos.com/calculator/4ddfrv7wvx`

[Further reading about parabolas on Wikipedia]
`https://en.wikipedia.org/wiki/Parabola`

6.7 Hyperbola

The *hyperbola* is another fundamental shape of nature.

Parameters

- F_1, F_2: the *focal points* of the hyperbola
- r_1: the distance from a point of the hyperbola to F_1
- r_2: the distance from a point of the hyperbola to F_2
- a: the semi-major axis of the hyperbola is the distance from the origin to the vertices V_1 and V_2
- b: the semi-minor axis of the hyperbola is the distance from a focus to the nearest asymptote
- c: the distance of the focal points from the centre. The distance between F_1 and F_2 is $2c$.
- ε: *eccentricity* of the hyperbola, $\varepsilon = \sqrt{1 + \frac{b^2}{a^2}} = \frac{c}{a}$

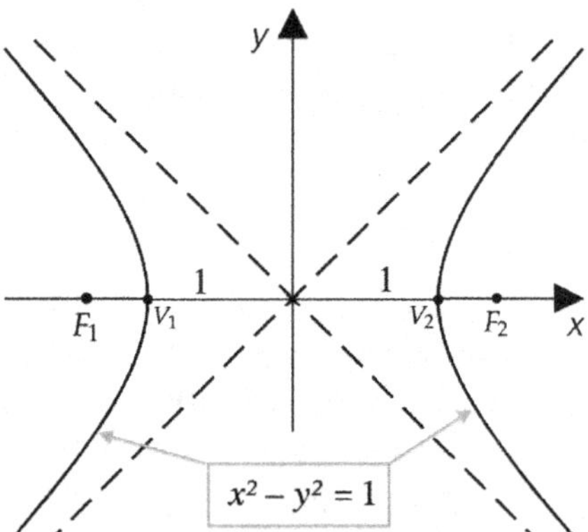

Figure 6.26: The graph of the unit hyperbola $x^2 - y^2 = 1$. The graph has two branches opening to the sides, and its eccentricity is $\varepsilon = \sqrt{1 + \frac{1}{1}} = \sqrt{2}$.

The graph of a hyperbola consists of two separate *branches*, as illustrated in Figure 6.26. The dashed lines are called the *asymptotes* of the hyperbola. The graph of the hyperbola approaches these lines but never touches them. The equations that describe these asymptotes are $y = \frac{b}{a}x$ and $y = -\frac{b}{a}x$.

Definition

A hyperbola is defined as the set of points such that the absolute value of the difference of the distances to the two focal points is constant:

$$|r_1 - r_2| = \text{const.}$$

Another way to define a hyperbola is as the set of points (x, y) that satisfy the equation

$$\frac{x^2}{a^2} - \frac{y^2}{b^2} = 1.$$

The coordinates of the two focal points of this hyperbola are

$$F_1 = (-c, 0) \qquad \text{and} \qquad F_2 = (c, 0),$$

where the *focal distance* is $c = \sqrt{a^2 + b^2}$. The coordinates of the *vertices* V_1 and V_2 are $(-a, 0)$ and $(a, 0)$.

The hyperbola's *eccentricity* is defined by the equation

$$\varepsilon = \sqrt{1 + \frac{b^2}{a^2}} = \frac{c}{a}.$$

The eccentricity is a number greater than 1 that determines the hyperbola's shape. Recall that an ellipse is also defined by an eccentricity parameter, though the formula is slightly different. This could be a coincidence—or is there a connection? Read on to find out.

Hyperbolic trigonometry

The study of the geometry of the points on the unit circle is called *circular trigonometry*. The geometry of the unit circle is described by the trigonometric functions $\sin\theta$ and $\cos\theta$. The function $\cos\theta$ defines the x-coordinates of the points on the unit circle, and $\sin\theta$ defines their y-coordinates. The point $P = (\cos\theta, \sin\theta)$ traces the unit circle as the angle θ goes from 0 to 2π.

Similarly, the study of the geometry of the unit hyperbola is called *hyperbolic trigonometry*. Doesn't that sound awesome? Next time your friends ask what you have been up to, tell them you are learning about hyperbolic trigonometry. Whereas we trace the path of the point P on the unit circle $x^2 + y^2 = 1$, we'll instead trace the path of a point Q on the unit hyperbola $x^2 - y^2 = 1$. We'll now define *hyperbolic* variants of the sin and cos functions to describe the coordinates of the point Q.

The coordinates of a point Q on the right branch of the unit hyperbola are $Q = (\cosh\mu, \sinh\mu)$, where μ is the *hyperbolic angle*. The x-coordinate of the point Q is $x = \cosh\mu$, and its y-coordinate is

$y = \sinh\mu$. The name hyperbolic angle is a bit of a misnomer, since μ actually measures an area. The area of the highlighted region in Figure 6.27 corresponds to $\frac{1}{2}\mu$.

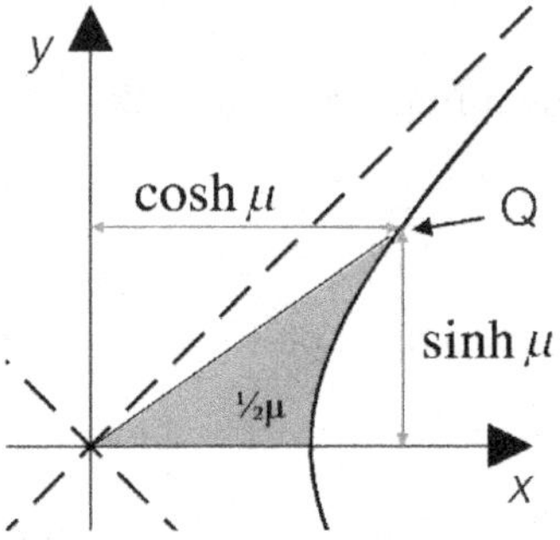

Figure 6.27: The functions $\cosh\mu$ and $\sinh\mu$ are defined as the x- and y-coordinates of a point moving on the unit hyperbola $x^2 - y^2 = 1$.

Recall the circular-trigonometric identity $\cos^2\theta + \sin^2\theta = 1$, which follows from the fact that all the points (x, y) on the unit circle obey $x^2 + y^2 = 1$. There is an analogous hyperbolic trigonometric identity:

$$\cosh^2\mu - \sinh^2\mu = 1.$$

This identity follows because we defined $x = \cosh\mu$ and $y = \sinh\mu$ to be the coordinates of a point Q which traces out the unit hyperbola $x^2 - y^2 = 1$.

The hyperbolic functions are related to the exponential function through the following formulas:

$$\cosh\mu = \frac{e^{\mu} + e^{-\mu}}{2}, \qquad \sinh\mu = \frac{e^{\mu} - e^{-\mu}}{2},$$

and

$$e^{\mu} = \cosh\mu + \sinh\mu.$$

Recall that even functions satisfy $f(-x) = f(x)$ and odd functions satisfy $g(-x) = -g(x)$. The cosh function is even, while sinh is odd. You can think of $\cosh x$ as the "even part" of e^x, and $\sinh x$ as the "odd part" of e^x.

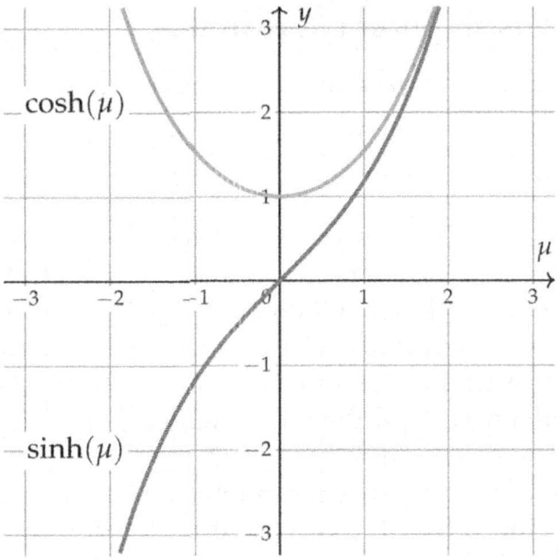

Figure 6.28: The graphs of the functions $\cosh\mu$ and $\sinh\mu$.

Don't worry about $\cosh\mu$ and $\sinh\mu$ too much. The hyperbolic trig functions are used much less often than the circular trigonometric functions $\cos\theta$ and $\sin\theta$. The main thing to remember is the general pattern: cosine functions are used to denote horizontal coordinates and sine functions are used to denote vertical coordinates.

The conic sections

There is a deep connection between the geometric shapes of the circle, the ellipse, the parabola, and the hyperbola. These seemingly different shapes can be obtained, geometrically speaking, from a single object: the cone. We can obtain the four curves by slicing the cone at different angles, as illustrated in Figure 6.29.

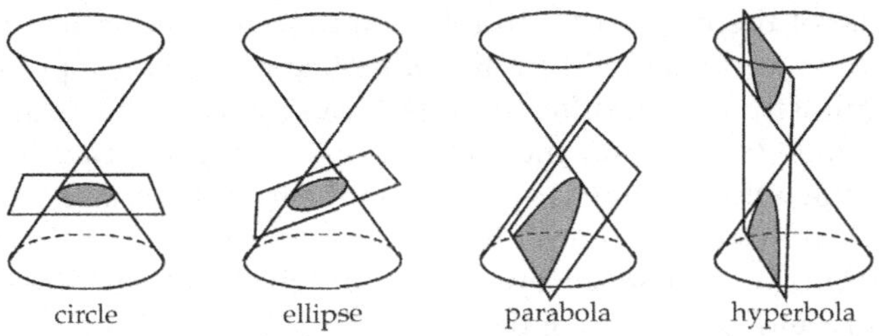

Figure 6.29: Taking slices through a cone at different angles produces different geometric shapes: a circle, an ellipse, a parabola, or a hyperbola.

Conic sections in polar coordinates

All four conic sections can be described by the same function in polar coordinates:

$$r(\theta) = \frac{q(1+\varepsilon)}{1+\varepsilon\cos(\theta)},$$

where q is the curve's closest distance to a focal point and ε is the curve's eccentricity. For a circle, $q = R$ (the radius) and the eccentricity parameter is $\varepsilon = 0$. For an ellipse, $q = a(1-\varepsilon)$ and the eccentricity parameter varies between 0 and 1 ($0 \leqslant \varepsilon < 1$). Note we include the case $\varepsilon = 0$ since a circle is a special case of an ellipse. For a parabola, $q = f$ (the focal length) and the eccentricity is $\varepsilon = 1$. For a hyperbola, $q = a(\varepsilon - 1)$ and the eccentricity is $\varepsilon > 1$.

We can use the eccentricity parameter ε to classify all four curves. Depending on the value of ε, the equation $r(\theta)$ defines either a circle, an ellipse, a parabola, or a hyperbola. Table 6.4 summarizes all our observations regarding conic sections.

Conic section	Equation	Polar function	Eccentricity
Circle	$x^2 + y^2 = R^2$	$r(\theta) = R$	$\varepsilon = 0$
Ellipse	$\frac{x^2}{a^2} + \frac{y^2}{b^2} = 1$	$r(\theta) = \frac{a(1-\varepsilon^2)}{1+\varepsilon\cos(\theta)}$	$\varepsilon = \sqrt{1-\frac{b^2}{a^2}},\ 0 \leqslant \varepsilon < 1$
Parabola	$y^2 = 4fx$	$r(\theta) = \frac{2f}{1+\cos(\theta)}$	$\varepsilon = 1$
Hyperbola	$\frac{x^2}{a^2} - \frac{y^2}{b^2} = 1$	$r(\theta) = \frac{a(\varepsilon^2-1)}{1+\varepsilon\cos(\theta)}$	$\varepsilon = \sqrt{1+\frac{b^2}{a^2}},\ 1 < \varepsilon < \infty$

Table 6.4: The four conic sections and their eccentricity parameters.

The motion of the planets is explained by Newton's law of gravitation. The gravitational interaction between two bodies always leads one of the two bodies to follow a trajectory described by one of the conic sections for which the other body is the focal point. Figure 6.30 illustrates four different trajectories for a satellite near planet F. The circle ($\varepsilon = 0$) and the ellipse ($0 \leqslant \varepsilon < 1$) describe *closed orbits,* in which the satellite is captured in the gravitational field of the planet F and remains in orbit forever. The parabola ($\varepsilon = 1$) and the hyperbola ($\varepsilon > 1$) describe *open orbits,* in which the satellite swings by the planet F and then continues.

Links

[Interactive graph of a hyperbola]
`https://www.desmos.com/calculator/2mnsk5o8vn`

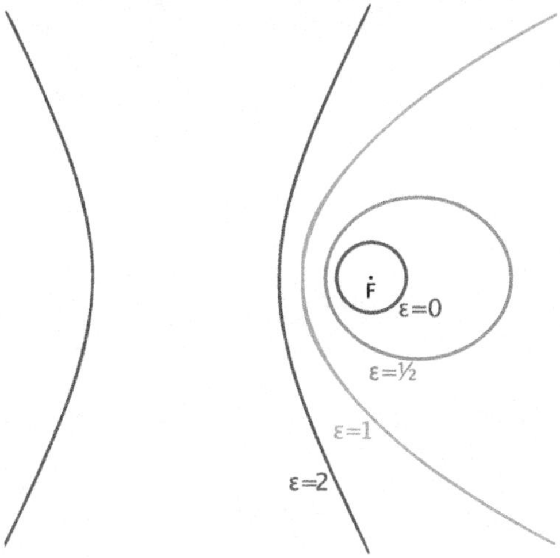

Figure 6.30: Four different trajectories for a satellite moving near a planet.

[Lots of information about conic sections on Wikipedia]
`https://en.wikipedia.org/wiki/Conic_section`
`http://en.wikipedia.org/wiki/Eccentricity_(mathematics)`

[An in-depth discussion on the conic sections]
`http://astrowww.phys.uvic.ca/~tatum/celmechs/celm2.pdf`

I'd love to continue this geometric digression and tell you more about the properties and applications of conic sections, but there are more pressing math topics to discuss! It's time to learn about vectors.

Chapter 7

Vectors

In this chapter we'll learn how to manipulate multi-dimensional objects called vectors. Vectors are the precise way to describe directions in space. We need vectors in order to describe physical quantities like forces, velocities, and accelerations.

Vectors are built from ordinary numbers, which form the *components* of the vector. You can think of a vector as a list of numbers, and *vector algebra* as operations performed on the numbers in the list. Vectors can also be manipulated as geometric objects, represented by arrows in space. For instance, the arrow that corresponds to the vector $\vec{v} = (v_x, v_y)$ starts at the origin $(0, 0)$ and ends at the point (v_x, v_y). The word vector comes from the Latin *vehere*, which means *to carry*. Indeed, the vector $\vec{v}$ takes the point $(0, 0)$ and carries it to the point (v_x, v_y).

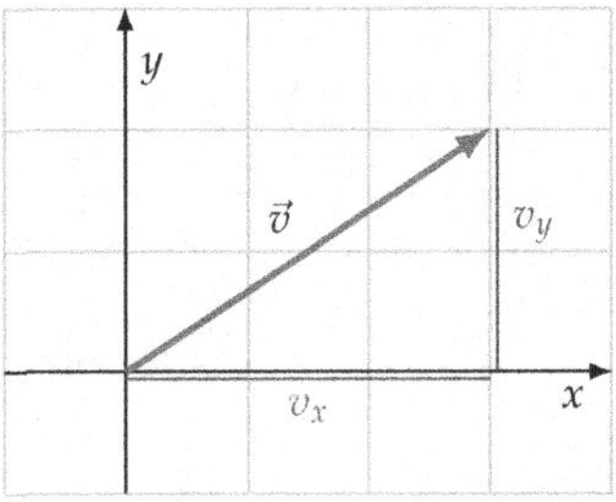

Figure 7.1: The vector $\vec{v} = (3, 2)$ is an arrow in the Cartesian plane. The horizontal component of $\vec{v}$ is $v_x = 3$ and the vertical component is $v_y = 2$.

This chapter will introduce you to vectors, vector algebra, and vector operations, which are very useful for solving physics problems. What you'll learn here applies more broadly to current problems in computer graphics, probability theory, machine learning, and other fields of science and mathematics. It's all about vectors these days, so you'd best get to know them.

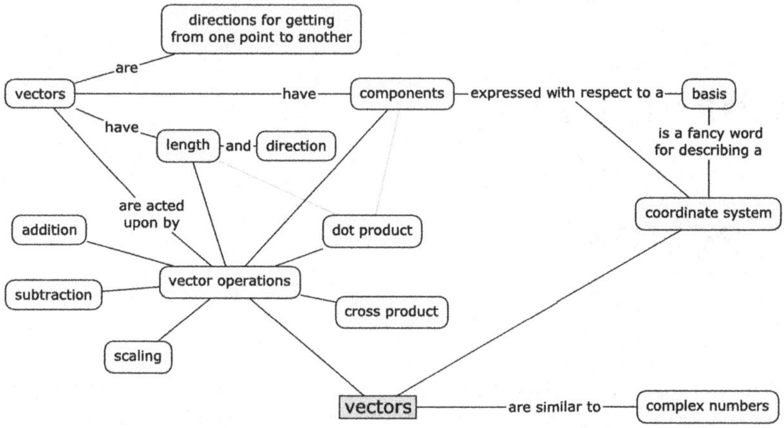

Figure 7.2: This figure illustrates the new concepts related to vectors. As you can see, there is quite a bit of new vocabulary to learn, but don't be fazed—all these terms are just fancy ways of talking about arrows.

7.1 Great outdoors

Vectors are directions for getting from point A to point B. Directions can be given in terms of street names and visual landmarks, or with respect to a coordinate system.

While on vacation in British Columbia, you want to visit a certain outdoor location your friend told you about. Your friend isn't available to take you there himself, but he has sent you *directions* for how to get to the place from the bus stop:

```
Sup G. Go to bus stop number 345. Bring a compass.
Walk 2 km north then 3 km east. You will find X there.
```

This text message contains all the information you need to find X.

Act 1: Following directions

You arrive at the bus stop, which is located at the top of a hill. From this height you can see the whole valley, and along the hillside below spreads a beautiful field of tall crops. The crops are so tall they prevent anyone standing in them from seeing too far; good thing you have a compass. You align the compass needle so the red arrow points north. You walk 2 km north, then turn 90° to the right so you're facing east, and walk another 3 km in that direction. You arrive at X as promised by your friend.

Okay, back to vectors. In this case, the *directions* can be also written as a vector $\vec{d}$, expressed as:

$$\vec{d} = 2\text{km}\,\hat{N} + 3\text{km}\,\hat{E}.$$

This is the mathematical expression that corresponds to the directions "Walk 2 km north then 3 km east." Here, $\hat{N}$ is a *direction* and the number in front of the direction tells you the distance to walk in that direction.

Act 2: Equivalent directions

Later during your vacation, you decide to return to the location X because you like the vegetation that grows there. You arrive at the bus stop to find there is a slight problem. From your position, you can see a kilometre to the north, where a group of armed and threatening-looking men stand, waiting to ambush anyone who tries to cross what has now become a trail through the crops. Clearly the word has spread about X and constant visitors have drawn too much attention to the location.

Well, technically speaking, there is no problem at X. The problem lies on the route that starts north and travels through the ambush squad. Can you find an alternate route that leads to X?

```
"Use math, Luke! Use math!"
```

Recall the commutative property of number addition: $a + b = b + a$. Maybe an analogous property holds for vectors? Indeed, it does:

$$\vec{d} = 2\text{km}\,\hat{N} + 3\text{km}\,\hat{E} = 3\text{km}\,\hat{E} + 2\text{km}\,\hat{N}.$$

The displacements in the $\hat{N}$ and $\hat{E}$ directions obey the commutative property. Since the directions can be followed in any order, you can first walk the 3 km east, then walk 2 km north and arrive at X again.

Act 3: Efficiency

It takes 5 km of walking to travel from the bus stop to X, and another 5 km to travel back to the bus stop. Thus, it takes a total of 10 km walking every time you want to go to X. Can you find a quicker route? What is the fastest way from the bus stop to the destination?

Instead of walking in the east and north directions, it would be quicker if you take the diagonal to the destination. Using Pythagoras' theorem you can calculate the length of the diagonal. When the side lengths are 3 and 2, the diagonal has length $\sqrt{3^2+2^2} = \sqrt{9+4} = \sqrt{13} = 3.60555\ldots$. The length of the diagonal route is just 3.6 km, which means the diagonal route saves you a whole 1.4 km of walking in each direction.

But perhaps seeking efficiency is not always necessary! You could take a longer path on the way back and give yourself time to enjoy the great outdoors.

Discussion

Vectors are directions for getting from one point to another point. To indicate directions on maps, we use the four cardinal directions: $\hat{N}$, $\hat{S}$, $\hat{E}$, $\hat{W}$. In math, however, we will use only two of the cardinals—$\hat{E} = \hat{x}$ and $\hat{N} = \hat{y}$—since they fit nicely with the usual way of drawing the Cartesian plane. We don't need an $\hat{S}$ direction because we can represent downward distances as negative distances in the $\hat{N}$ direction. Similarly, $\hat{W}$ is the same as negative $\hat{E}$.

From now on, when we talk about vectors we will always represent them with respect to the standard coordinate system $\hat{x}$ and $\hat{y}$, and use *bracket notation*,

$$(v_x, v_y) = v_x\,\hat{x} \;+\; v_y\,\hat{y}.$$

Bracket notation is nice because it's compact, which is good since we will be doing a lot of calculations with vectors. Instead of explicitly writing out all the directions, we will automatically assume that the first number in the bracket is the $\hat{x}$ distance and the second number is the $\hat{y}$ distance.

7.2 Vectors

Vectors are extremely useful in all areas of life. In physics, for example, we use a vector to describe the velocity of an object. It is not sufficient to say that the speed of a tennis ball is 200 kilometres per hour: we must also specify the direction in which the ball is moving. Both of the two velocities

$$\vec{v}_1 = (200, 0) \qquad \text{and} \qquad \vec{v}_2 = (0, 200)$$

describe motion at the speed of 200 kilometres per hour; but since one velocity points along the x-axis, and the other points along the y-axis, they are *completely* different velocities. The velocity vector contains information about the object's speed *and* its direction. The direction makes a big difference. If it turns out the tennis ball is hurtling toward you, you'd better get out of the way!

The main idea in this chapter is that **vectors are not the same as numbers**. We'll start by defining what vectors are. Then we'll describe all the mathematical operations we can perform with vectors, which include vector addition $\vec{u}+\vec{v}$, vector subtraction $\vec{u}-\vec{v}$, vector scaling $\alpha\vec{v}$, and other operations. In Section 7.4 we'll also talk about two different kinds of vector products.

Definitions

A two-dimensional vector $\vec{v}$ corresponds to a *pair of numbers*:

$$\vec{v} = (v_x, v_y),$$

where v_x is the *x-component* of the vector and v_y is its *y-component*. We denote the set of two-dimensional vectors as $\mathbb{R}^2$, since the components of a two-dimensional vector are specified by two real numbers. We'll use the mathematical shorthand $\vec{v} \in \mathbb{R}^2$ to define a two-dimensional vector $\vec{v}$. Vectors in $\mathbb{R}^2$ can be represented as arrows in the Cartesian plane. See the vector $\vec{v} = (3,2)$ illustrated in Figure 7.1.

We can also define three-dimensional vectors like the vector $\vec{v} = (v_x, v_y, v_z) \in \mathbb{R}^3$, which has three components. Three-dimensional vectors can be represented as arrows in a coordinate system that has three axes, like the one shown in Figure 7.10 on page 146. A three-dimensional coordinate system is similar to the Cartesian coordinate system you're familiar with, and includes the additional z-axis that measures the height above the plane. In fact, there's no limit to the number of dimensions for vectors. We can define vectors in an n-dimensional space: $\vec{v} = (v_1, v_2, \ldots, v_n) \in \mathbb{R}^n$. For the sake of simplicity, we'll define all the vector operation formulas using two-dimensional vectors. Unless otherwise indicated in the text, all the formulas we give for two-dimensional vectors $\vec{v} \in \mathbb{R}^2$ also apply to n-dimensional vectors $\vec{v} \in \mathbb{R}^n$.

Vector operations

Consider two vectors, $\vec{u} = (u_x, u_y)$ and $\vec{v} = (v_x, v_y)$, and assume that $\alpha \in \mathbb{R}$ is an arbitrary constant. The following operations are defined for these vectors:

- **Addition:** $\vec{u}+\vec{v} = (u_x + v_x,\ u_y + v_y)$
- **Subtraction:** $\vec{u}-\vec{v} = (u_x - v_x,\ u_y - v_y)$
- **Scaling:** $\alpha\vec{u} = (\alpha u_x,\ \alpha u_y)$
- **Dot product:** $\vec{u}\cdot\vec{v} = u_x v_x + u_y v_y$

- **Length:** $\|\vec{u}\| = \sqrt{\vec{u}\cdot\vec{u}} = \sqrt{u_x^2 + u_y^2}$. The vector's length is also called the *norm* of the vector. We sometimes use the letter u to denote the length of the vector $\vec{u}$.

Note there is no vector division operation.

For vectors in a three-dimensional space $\vec{u} = (u_x, u_y, u_z) \in \mathbb{R}^3$ and $\vec{v} = (v_x, v_y, v_z) \in \mathbb{R}^3$, we can also define the **cross product** operation $\vec{u} \times \vec{v} = (u_y v_z - u_z v_y,\ u_z v_x - u_x v_z,\ u_x v_y - u_y v_x)$. The dot product and the cross product are new operations that you probably haven't seen before. We'll talk more about dot products and the cross products in Section 7.4. For now let's start with the basics.

Vector representations

We'll use three equivalent ways to denote vectors in two dimensions:

- $\vec{v} = (v_x, v_y)$: component notation. The vector is written as a pair of numbers called the *components* or *coordinates* of the vector.
- $\vec{v} = v_x\hat{\imath} + v_y\hat{\jmath}$: unit vector notation. The vector is expressed as a combination of the unit vectors $\hat{\imath} = (1,0)$ and $\hat{\jmath} = (0,1)$.
- $\vec{v} = \|\vec{v}\|\angle\theta$: length-and-direction notation (polar coordinates). The vector is expressed in terms of its *length* $\|\vec{v}\|$ and the angle θ that the vector makes with the x-axis.

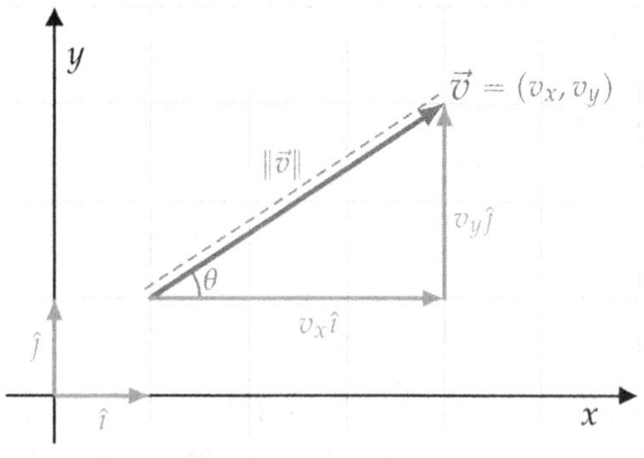

Figure 7.3: The vector $\vec{v} = (v_x, v_y) = v_x\hat{\imath} + v_y\hat{\jmath} = \|\vec{v}\|\angle\theta$.

We use the component notation for doing vector algebra calculations since it is most compact. The unit vector notation shows explicitly that the vector $\vec{v}$ corresponds to the sum of $v_x\hat{\imath}$ (a displacement of v_x steps in the direction of the x-axis) and $v_y\hat{\jmath}$ (a displacement of v_y steps in the direction of the y-axis). The length-and-direction notation describes the vector $\vec{v}$ as a displacement of $\|\vec{v}\|$ steps in the direction of the angle θ. We'll use all three ways of denoting vectors throughout the rest of the book, and we'll learn how to convert between them.

Vector algebra

Addition and subtraction Just like numbers, you can add vectors

$$\vec{v} + \vec{w} = (v_x, v_y) + (w_x, w_y) = (v_x + w_x, v_y + w_y),$$

subtract them

$$\vec{v} - \vec{w} = (v_x, v_y) - (w_x, w_y) = (v_x - w_x, v_y - w_y),$$

and solve all kinds of equations where the unknown variable is a vector. This is not a formidably complicated new development in mathematics. Performing arithmetic calculations on vectors simply requires **carrying out arithmetic operations on their components**. Given two vectors, $\vec{v} = (4,2)$ and $\vec{w} = (3,7)$, their difference is computed as $\vec{v} - \vec{w} = (4,2) - (3,7) = (1,-5)$.

Scaling We can also *scale* a vector by any number $\alpha \in \mathbb{R}$:

$$\alpha\vec{v} = (\alpha v_x, \alpha v_y),$$

where each component is multiplied by the scale factor α. Scaling changes the length of a vector. If $\alpha > 1$ the vector will get longer, and if $0 \leqslant \alpha < 1$ then the vector will become shorter. If α is a negative number, the scaled vector will point in the opposite direction.

Length A vector's length is obtained from Pythagoras' theorem. Imagine a right-angle triangle with one side of length v_x and the other side of length v_y; the length of the vector is equal to the length of the triangle's hypotenuse:

$$\|\vec{v}\|^2 = v_x^2 + v_y^2 \qquad \Rightarrow \qquad \|\vec{v}\| = \sqrt{v_x^2 + v_y^2}\,.$$

A common technique is to scale a vector $\vec{v}$ by one over its length $\frac{1}{\|\vec{v}\|}$ to obtain a unit vector that points in the same direction as $\vec{v}$:

$$\hat{v} = \frac{\vec{v}}{\|\vec{v}\|} = \left(\frac{v_x}{\|\vec{v}\|}, \frac{v_y}{\|\vec{v}\|} \right).$$

Unit vectors (denoted with a hat instead of an arrow) are useful when you want to describe only a direction in space without any specific length in mind. Verify that $\|\hat{v}\| = 1$.

Vector as arrows

So far, we described how to perform algebraic operations on vectors in terms of their components. Vector operations can also be interpreted geometrically, as operations on arrows in the Cartesian plane.

Vector addition The sum of two vectors corresponds to the combined displacement of the two vectors. Figure 7.4 illustrates the addition of two vectors, $\vec{v}_1 = (3,0)$ and $\vec{v}_2 = (2,2)$. The sum of the two vectors is the vector $\vec{v}_1 + \vec{v}_2 = (3,0) + (2,2) = (5,2)$.

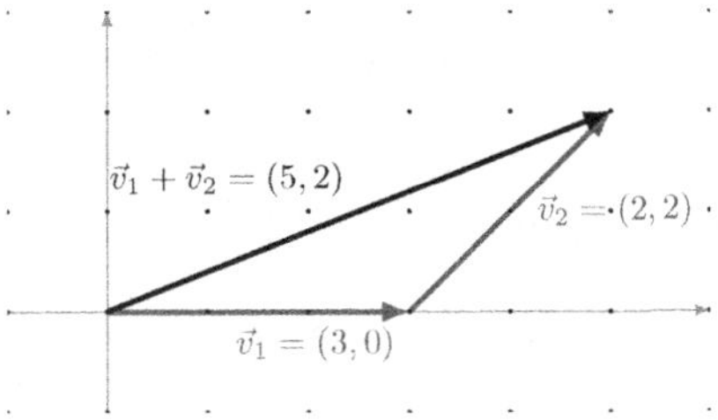

Figure 7.4: The addition of the vectors $\vec{v}_1$ and $\vec{v}_2$ produces the vector $(5,2)$.

Vector subtraction Before we describe vector subtraction, note that multiplying a vector by a scale factor $\alpha = -1$ gives a vector of the same length as the original, but pointing in the opposite direction.

This fact is useful if you want to subtract two vectors using the graphical approach. Subtracting a vector is the same as adding the negative of the vector:

$$\vec{w} - \vec{v}_1 = \vec{w} + (-\vec{v}_1) = \vec{v}_2.$$

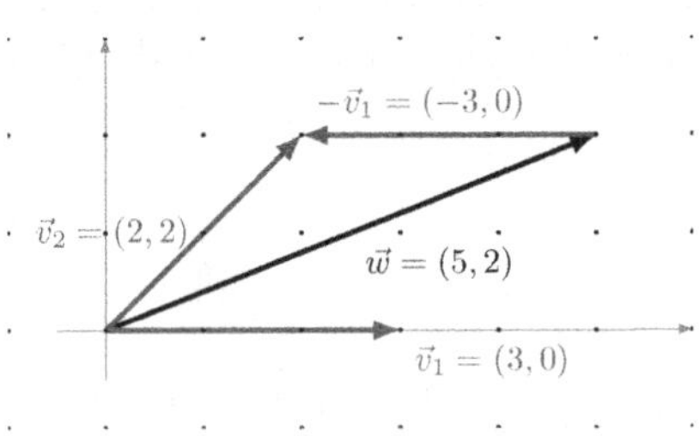

Figure 7.5: The vector subtraction $\vec{w} - \vec{v}_1$ is equivalent to the vector addition $\vec{w} + (-\vec{v}_1)$, where $(-\vec{v}_1)$ is like $\vec{v}_1$ but points in the opposite direction.

Figure 7.5 illustrates the graphical procedure for subtracting the vector $\vec{v}_1 = (3,0)$ from the vector $\vec{w} = (5,2)$. Subtraction of $\vec{v}_1 = (3,0)$ is the same as addition of $-\vec{v}_1 = (-3,0)$.

Scaling The scaling operation acts to change the length of a vector. Suppose we want to obtain a vector in the same direction as the vector $\vec{v} = (3,2)$, but half as long. "Half as long" corresponds to a scale factor of $\alpha = 0.5$. The scaled-down vector is $\vec{w} = 0.5\vec{v} = (1.5,1)$. Conversely, we can think of the vector $\vec{v}$ as being twice as long as the vector $\vec{w}$.

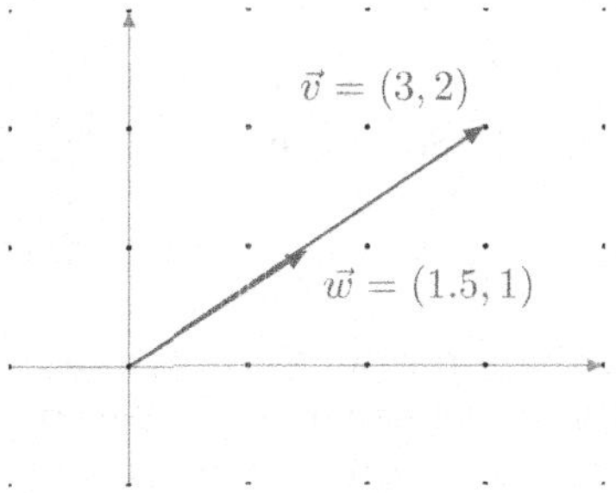

Figure 7.6: Vectors $\vec{v}$ and $\vec{w}$ are related by the equation $\vec{v} = 2\vec{w}$.

Multiplying a vector by a negative number reverses its direction.

Length-and-direction representation

So far, we've seen how to represent a vector in terms of its components. There is another way of representing two-dimensional vectors: we can describe the vector $\vec{v} \in \mathbb{R}^2$ in terms of its length $\|\vec{v}\|$ and its direction θ—the angle it makes with the x-axis. For example, the vector $(1,1)$ can also be written as $\sqrt{2}\angle 45°$ in polar coordinates. This length-and-direction notation is useful because it makes it easy to see the "size" of vectors. On the other hand, vector arithmetic operations are much easier to carry out in the component notation. It's therefore good to know the formulas for converting between the two vector representations.

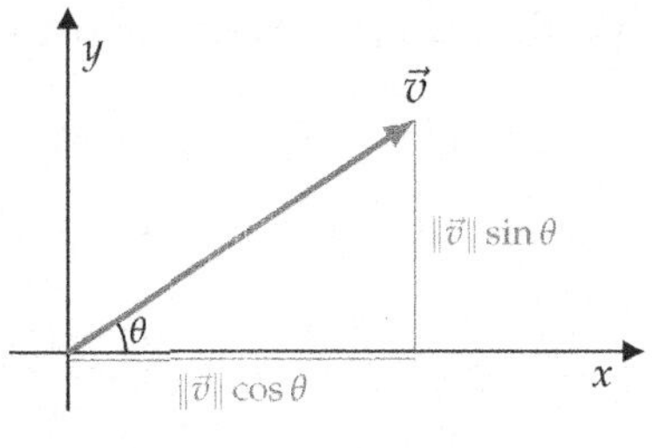

Figure 7.7: The x- and y-components of a vector with length $\|\vec{v}\|$ in the direction θ are given by $v_x = \|\vec{v}\| \cos\theta$ and $v_y = \|\vec{v}\| \sin\theta$.

To convert the length-and-direction vector $\vec{v} = \|\vec{v}\|\angle\theta$ into an x-component and a y-component (v_x, v_y), use the formulas

$$v_x = \|\vec{v}\|\cos\theta \quad \text{and} \quad v_y = \|\vec{v}\|\sin\theta.$$

To convert from component notation $\vec{v} = (v_x, v_y)$ to length-and-direction $\|\vec{v}\|\angle\theta$, use

$$\|\vec{v}\| = \sqrt{v_x^2 + v_y^2}, \quad \theta = \begin{cases} \tan^{-1}\left(\frac{v_y}{v_x}\right) & \text{if } v_x > 0, \\ 180^\circ + \tan^{-1}\left(\frac{v_y}{v_x}\right) & \text{if } v_x < 0, \\ 90^\circ & \text{if } v_x = 0 \text{ and } v_y > 0, \\ -90^\circ & \text{if } v_x = 0 \text{ and } v_y < 0. \end{cases}$$

We've already seen these formulas in Section 6.4 (page 111), when we learned about the transformations between Cartesian and polar coordinates for points. The conversion procedure for vectors is exactly the same, including the trickiness around calculating θ when v_x is negative or zero. I invite you to revisit exercise E6.8 on page 115 to review the conversion operations between Cartesian coordinates and polar coordinates.

Unit vector notation

In two dimensions, we can think of a vector $\vec{v} = (v_x, v_y)$ as a command to "Go a distance v_x in the x-direction and a distance v_y in the y-direction." To write this set of commands more explicitly, we can use multiples of the vectors $\hat{\imath}$ and $\hat{\jmath}$. These are the unit vectors pointing in the x and y directions:

$$\hat{\imath} = (1,0) \qquad \text{and} \qquad \hat{\jmath} = (0,1).$$

Any number multiplied by $\hat{\imath}$ corresponds to a vector with that number in the first coordinate. For example, $3\hat{\imath} = (3,0)$ and $4\hat{\jmath} = (0,4)$.

In physics, we tend to perform a lot of numerical calculations with vectors; to make things easier, we often use unit vector notation:

$$v_x\hat{\imath} + v_y\hat{\jmath} \qquad \Leftrightarrow \qquad (v_x, v_y).$$

The addition rule remains the same for the new notation:

$$\underbrace{2\hat{\imath} + 3\hat{\jmath}}_{\vec{v}} + \underbrace{5\hat{\imath} - 2\hat{\jmath}}_{\vec{w}} = \underbrace{7\hat{\imath} + 1\hat{\jmath}}_{\vec{v}+\vec{w}}.$$

It's the same story repeating all over again: we need to add $\hat{\imath}$s with $\hat{\imath}$s, and $\hat{\jmath}$s with $\hat{\jmath}$s.

Examples

Simple example

Compute the sum $\vec{s} = 4\hat{\imath} + 5\angle 30^\circ$. Express your answer in the length-and-direction notation.

Since we want to carry out an addition, and since addition is performed in terms of components, our first step is to convert $5\angle 30^\circ$ into component notation: $5\angle 30^\circ = 5\cos 30^\circ\hat{\imath} + 5\sin 30^\circ\hat{\jmath} = \frac{5\sqrt{3}}{2}\hat{\imath} + \frac{5}{2}\hat{\jmath}$. We can now compute the sum:

$$\vec{s} \;=\; 4\hat{\imath} \;+\; \tfrac{5\sqrt{3}}{2}\hat{\imath} + \tfrac{5}{2}\hat{\jmath} \;=\; (4 + \tfrac{5\sqrt{3}}{2})\hat{\imath} + (\tfrac{5}{2})\hat{\jmath}.$$

The x-component of the sum is $s_x = (4 + \frac{5\sqrt{3}}{2})$ and the y-component of the sum is $s_y = (\frac{5}{2})$. To express the answer as a length and a direction, we compute the length $\|\vec{s}\| = \sqrt{s_x^2 + s_y^2} = 8.697$ and the direction $\tan^{-1}(s_y/s_x) = 16.7^\circ$. The answer is $\vec{s} = 8.697\angle 16.7^\circ$.

Vector addition example

You're heading to physics class after a "safety meeting" with a friend, and are looking forward to two hours of finding absolute amazement and awe in the laws of Mother Nature. As it turns out, there is no enlightenment to be had that day because there is going to be an in-class midterm. The first question involves a block sliding down an incline. You look at it, draw a little diagram, and then wonder how the hell you are going to find the net force acting on the block. The three forces acting on the block are $\vec{W} = 300\angle -90^\circ$, $\vec{N} = 260\angle 120^\circ$, and $\vec{F}_f = 50\angle 30^\circ$.

You happen to remember the net force formula:

$$\sum \vec{F} = \vec{F}_{\text{net}} = m\vec{a} \qquad [\text{ Newton's } 2^{\text{nd}} \text{ law }].$$

You get the feeling Newton's 2nd law is the answer to all your troubles. You sense this formula is certainly the key because you saw the keyword "net force" when reading the question, and notice "net force" also appears in this very equation.

The net force is the sum of all forces acting on the block:

$$\vec{F}_{\text{net}} = \sum \vec{F} = \vec{W} + \vec{N} + \vec{F}_f.$$

All that separates you from the answer is the addition of these vectors. Vectors have components, and there is the whole sin/cos procedure for decomposing length-and-direction vectors into their components. If you have the vectors as components you'll be able to add them and find the net force.

Okay, chill! Let's do this one step at a time. The net force must have an x-component, which, according to the equation, must equal the sum of the x-components of all the forces:

$$\begin{aligned} F_{\text{net},x} &= W_x + N_x + F_{f,x} \\ &= 300\cos(-90°) + 260\cos(120°) + 50\cos(30°) \\ &= -86.7. \end{aligned}$$

Now find the y-component of the net force using the sin of the angles:

$$\begin{aligned} F_{\text{net},y} &= W_y + N_y + F_{f,y} \\ &= 300\sin(-90°) + 260\sin(120°) + 50\sin(30°) \\ &= -49.8. \end{aligned}$$

Combining the two components of the vector, you get the final answer:

$$\begin{aligned} \vec{F}_{\text{net}} &= (F_{\text{net},x}, F_{\text{net},y}) \\ &= (-86.7, -49.8) = -86.7\hat{\imath} - 49.8\hat{\jmath} \\ &= 100\angle 209.9°, \end{aligned}$$

where you found the angle 209.9° by computing $\tan^{-1}(49.8/86.7)$ and adding 180° since the x-component is negative. Bam! Just like that you're done, because you overstand them vectors!

Relative motion example

A boat can reach a top speed of 12 knots in calm seas. Instead of cruising through a calm sea, however, the boat's crew is trying to sail up the St-Laurence river. The speed of the current is 5 knots.

If the boat travels directly upstream at full throttle $12\hat{\imath}$, then the speed of the boat relative to the shore will be

$$12\hat{\imath} - 5\hat{\imath} = 7\hat{\imath},$$

since we must "deduct" the speed of the current from the speed of the boat relative to the water. See the vector diagram in Figure 7.8.

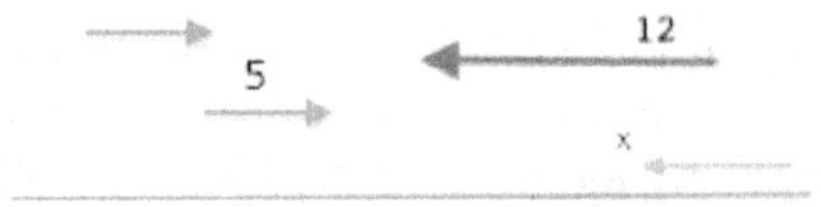

Figure 7.8: A boat travels with speed 12 knots against a current of 5 knots.

If the crew wants to cross the river perpendicular to the current flow, they can use some of the boat's thrust to counterbalance the current, and the remaining thrust to push across. The situation is illustrated in Figure 7.9. In what direction should the boat sail to cross the river? We are looking for the direction of $\vec{v}$ the boat should take such that, after adding in the velocity of the current, the boat moves in a straight line between the two banks (in the $\hat{\jmath}$ direction).

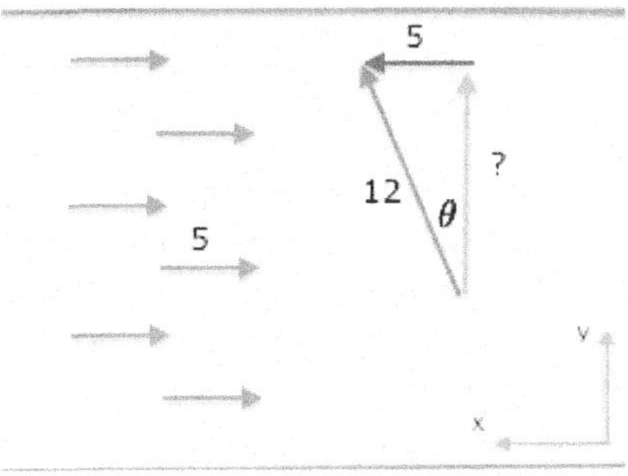

Figure 7.9: Part of the boat's thrust cancels the current.

Let's analyze the vector diagram. The opposite side of the triangle is parallel to the current flow and has length 5. We take the up-the-river component of the velocity $\vec{v}$ to be equal to $5\hat{\imath}$, so that it cancels exactly the $-5\hat{\imath}$ flow of the river. The hypotenuse has length 12 since this is the speed of the boat relative to the surface of the water.

From all of this we can answer the question like professionals. You want the angle? Well, we have that $\frac{\text{opp}}{\text{hyp}} = \frac{5}{12} = \sin(\theta)$, where θ is the angle of the boat's course relative to the straight line between the two banks. We can use the inverse-sin function to solve for the angle:

$$\theta = \sin^{-1}\left(\tfrac{5}{12}\right) = 24.62^\circ.$$

The across-the-river component of the velocity can be calculated using $v_y = 12\cos(\theta) = 10.91$, or from Pythagoras' theorem if you prefer $v_y = \sqrt{\|\vec{v}\|^2 - v_x^2} = \sqrt{12^2 - 5^2} = 10.91$.

Discussion

We did a lot of hands-on activities with vectors in this section and skipped over some of the theoretical details. Now that you've been exposed to the practical side of vector calculations, it's worth clarifying certain points that we glossed over.

Vectors vs. points

We used the notation $\mathbb{R}^2$ to describe two kinds of math objects: the set of points in the Cartesian plane and the set of vectors in a two-dimensional space. The point $P = (P_x, P_y)$ and the vector $\vec{v} = (v_x, v_y)$ are both represented by pairs of real numbers, so we use the notation $P \in \mathbb{R}^2$ and $\vec{v} \in \mathbb{R}^2$ to describe them. This means that a pair of numbers $(3, 2) \in \mathbb{R}^2$ could represent the *coordinates* of a point, or the *components* of a vector, depending on the context.

Let's take a moment to review the definitions of points and vectors and clarify the types of operations we can perform on them:

- **Space of points** $\mathbb{R}^2$: the set of points $P = (P_x, P_y)$ corresponds to locations in the Cartesian plane. The point $P = (P_x, P_y)$ corresponds to the geometric instructions: "Starting at the origin $(0, 0)$, move P_x units along the x-axis and P_y units along the y-axis." The distance between points P and Q is denoted $d(P, Q)$.
- **Vector space** $\mathbb{R}^2$: the set of vectors $\vec{v} = (v_x, v_y)$ describes displacements in the Cartesian plane. The vector $\vec{v} = (v_x, v_y)$ corresponds to the instructions: "Starting anywhere, move v_x units along the x-axis and v_y units along the y-axis." Vectors can be combined and manipulated using the vector algebra operations $\vec{u} + \vec{v}$, $\vec{u} - \vec{v}$, $\alpha\vec{u}$, $\vec{u} \cdot \vec{v}$, and $\|\vec{v}\|$.

Note the geometric instructions for points and vectors are very similar; the only difference is the starting point. The coordinates of a point (P_x, P_y) specify a *fixed position* relative to the origin $(0, 0)$, while the components of a vector (v_x, v_y) describe a *relative displacement* that can have any starting point.

Let's look at some examples of calculations that combine points and vectors. Consider the points P and Q in the Cartesian plane, and the displacement vector $\vec{v}_{PQ}$ between them. The displacement vector $\vec{v}_{PQ}$ gives the "move instructions" for getting from point P to point Q and is defined by the equation:

$$\vec{v}_{PQ} = Q - P.$$

This equation says that subtracting two points produces a vector, which make sense if you think about it—the "difference" between two points is a displacement vector.

We can use the displacement vector $\vec{v}_{PQ}$ in calculations like this:

$$P + \vec{v}_{PQ} = P + (Q - P) = Q.$$

In words, this calculation shows that "Starting at the point P and moving by $\vec{v}_{PQ}$ brings us to the point Q."

The above equations use addition and subtraction operations between a mix of points and vectors. This is rather unusual: normally we only use operations like "+" and "-" between math objects of the same kind. In this case, we're allowed to mix points and vectors because they both describe "move instructions" of the same kind.

Let's keep going. What other useful calculations can we do by combining points and vectors? Suppose we wanted to find the midpoint M that lies exactly in the middle between points P and Q. We can find the midpoint M using the displacement vector $\vec{v}_{PQ}$ and some basic vector algebra. If starting from P and moving by $\vec{v}_{PQ}$ brings us all the way to the point Q, then starting from P and moving by $\frac{1}{2}\vec{v}_{PQ}$ will bring us to the midpoint: $M = P + \frac{1}{2}\vec{v}_{PQ}$.

The mathematical bridge between points and vectors allows us to use vector techniques to solve geometry problems. By learning to describe geometric objects like points, lines, and circles using vectors, we can do complicated geometry calculations using simple algebraic manipulations like vector operations. This exemplifies a general pattern in mathematics: applying techniques developed in one domain to solve problems in another domain.

Example You come to class one day and there's a surprise quiz that asks you to write the formula for the distance $d(P, Q)$ between two points $P = (P_x, P_y)$ and $Q = (Q_x, Q_y)$. You don't remember ever learning about such a formula and feel caught off guard. How can the teacher ask for a formula they haven't covered in class yet? This seems totally unfair!

After a minute of stressing out, you take a deep breath, come back to your senses, and resolve to give this problem a shot. You start by sketching a coordinate system, placing points P and Q in it, and drawing the line that connects the two points. What is the formula that describes the length of this line?

The line from P to Q looks like the hypotenuse of a triangle, which makes you think that trigonometry could somehow be used to find the answer. Unfortunately, trying to remember the trigonometry formulas has only the effect of increasing your math anxiety. You take this as a sign that you should look for other options. In math, it's important to trust your gut instincts.

By a fortunate coincidence, you were recently reading about the connection between points and vectors, and specifically about the displacement vector $\vec{v}_{PQ} = Q - P$. The line in your sketch represents the vector $\vec{v}_{PQ}$. You realize that the distance between the points P and Q is the same as the length of the vector $\vec{v}_{PQ}$. You remember the formula for the length of a vector $\vec{v}$ is $\|\vec{v}\| = \sqrt{v_x^2 + v_y^2}$

and you know the formula for the displacement vector is $\vec{v}_{PQ} = (Q_x - P_x, Q_y - P_y)$, so you combine these formulas to obtain the answer: $d(P,Q) = \|\vec{v}_{PQ}\| = \sqrt{(Q_x - P_x)^2 + (Q_y - P_y)^2}$. One more win for the "don't worry and try it" strategy for solving math problems!

Vectors in three dimensions

A three-dimensional coordinate system consists of three axes: the x-axis, the y-axis, and the z-axis. The three axes point in perpendicular directions to each other, as illustrated in Figure 7.10. Look around you and find a corner of the room you're in where two walls and the floor meet. The x-axis and the y-axis are the edges where the floor meets the walls. The vertical edge where the two walls meet represents the z-axis.

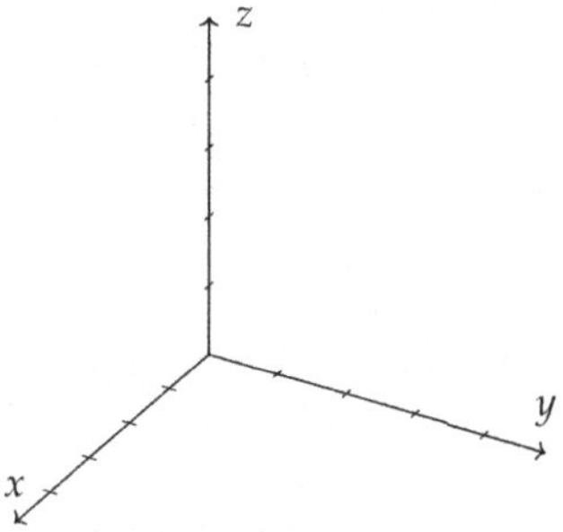

Figure 7.10: A three-dimensional coordinate system with x, y, and z axes.

The vector $\vec{v} = (v_x, v_y, v_z) \in \mathbb{R}^3$ describes the following displacement instructions: "Move v_x units in the direction of the x-axis, then move v_y along the y-axis, and finally move v_z in the direction of the z-axis." In three dimensions, there are three unit vectors that describe unit steps in the direction of each of the axes:

$$\hat{\imath} = (1,0,0), \qquad \hat{\jmath} = (0,1,0) \quad \text{and} \quad \hat{k} = (0,0,1).$$

We can therefore describe the vector $\vec{v} = (v_x, v_y, v_z)$ in terms of unit vectors as $\vec{v} = v_x\hat{\imath} + v_y\hat{\jmath} + v_z\hat{k}$.

High-dimensional vectors

The most common types of vectors you'll encounter in math and physics are two-dimensional and three-dimensional vectors. In other fields of science like genetics and machine learning, it's common to see vectors with many more dimensions. For example, in machine learning we often represent "rich data" like images, videos, and text as vectors with thousands of dimensions.

An example of an n-dimensional vector is

$$\vec{v} = (v_1, v_2, \ldots, v_n) \in \mathbb{R}^n.$$

The vector algebra operations you learned in this section also apply to these high-dimensional vectors.

Vectors and vector coordinates

One final point we need to clarify is the difference between real-world vector quantities like the velocity of a tennis ball $\vec{v}$ and its mathematical representation as a coordinate vector (v_x, v_y, v_z). If you know the coordinate vector (v_x, v_y, v_z) then you know what the real-world velocity is, right? Not quite.

Let's say you're doing a physics research project on tennis serves. You define an xyz-coordinate system for the tennis court, which allows you to represent the ball's velocity $\vec{v}$ as a triple of components (v_x, v_y, v_z) interpreted as: "The ball is moving with velocity v_x units in the x-direction, v_y units in the y-direction, and v_z units in the z-direction."

Suppose you want to describe the velocity vector $\vec{v}$ to a fellow physicist via text message. Referring to your sheet of calculations, you find the values $\vec{v} = (60, 3, -2)$, which you know were measured in metres per second. You send this message:

```
The velocity is (60,3,-2) measured in metres per second.
```

A few minutes later the following reply comes back:

```
Wait whaaat? What coordinate system are you using?
```

Indeed the information you sent is incomplete. Vector components depend on the coordinate system in which the vectors are represented. The triple of numbers $(60, 3, -2)$ only makes sense once you know the directions of the axes in the xyz-coordinate system. Realizing your mistake, you send a text with all the required information:

```
Using the coordinate system centred at the south post of
the net, with the x-axis pointing east along the court,
the y-axis pointing north along the net, and the z-axis
pointing up, the velocity is (60,3,-2) in metres per second.
```

A few seconds later, you get the reply:

```
OK got it now. Thx!
```

This hypothetical situation illustrates the importance of the coordinate systems for describing vectors. If you don't know what the coordinate system is, knowing the coordinates (v_x, v_y, v_z) doesn't tell you much. Only when you know the directions of the unit vectors $\hat{\imath}$, $\hat{\jmath}$, and $\hat{k}$ can you interpret the instructions $\vec{v} = v_x\hat{\imath} + v_y\hat{\jmath} + v_z\hat{k}$.

It turns out, using the xyz-coordinate system with the three vectors $\{\hat{\imath}, \hat{\jmath}, \hat{k}\}$ is just one of many possible ways we can represent vectors. We can represent a vector $\vec{v}$ as coordinates (v_1, v_2, v_3) with respect to any *basis* $\{\hat{e}_1, \hat{e}_2, \hat{e}_3\}$ using the expression $\vec{v} = v_1\hat{e}_1 + v_2\hat{e}_2 + v_3\hat{e}_3$, which corresponds to the instructions: "Move v_1 units in the direction of $\hat{e}_1$, move v_2 units in the direction of $\hat{e}_2$, and move v_3 units in the direction of $\hat{e}_3$."

What's a basis, you ask? I'm glad you asked, because this is the subject of the next section.

7.3 Basis

One of the most important concepts in the study of vectors is the concept of a *basis*. Consider the three-dimensional vector space $\mathbb{R}^3$. A *basis* for $\mathbb{R}^3$ is a set of vectors $\{\hat{e}_1, \hat{e}_2, \hat{e}_3\}$ that can be used as a coordinate system for $\mathbb{R}^3$. If the set of vectors $\{\hat{e}_1, \hat{e}_2, \hat{e}_3\}$ is a basis, then you can *represent* any vector $\vec{v} \in \mathbb{R}^3$ as coordinates (v_1, v_2, v_3) *with respect to* that basis:

$$\vec{v} = v_1\hat{e}_1 + v_2\hat{e}_2 + v_3\hat{e}_3.$$

The vector $\vec{v}$ is obtained by measuring out a distance v_1 in the $\hat{e}_1$ direction, a distance v_2 in the $\hat{e}_2$ direction, and a distance v_3 in the $\hat{e}_3$ direction.

You are already familiar with the *standard* basis $\{\hat{\imath}, \hat{\jmath}, \hat{k}\}$, which is associated with the xyz-coordinate system. You know that any vector $\vec{v} \in \mathbb{R}^3$ can be expressed as a triple (v_x, v_y, v_z) with respect to the basis $\{\hat{\imath}, \hat{\jmath}, \hat{k}\}$ through the formula $\vec{v} = v_x\hat{\imath} + v_y\hat{\jmath} + v_z\hat{k}$. The whole point of this section is to let you know that other bases (coordinate systems) exist, and to get you into the habit of asking, "With respect to which coordinate system?" every time you see a coordinate vector (a, b, c).

An analogy

Let's start with a simple example of a basis. If you look at the HTML source code behind any web page, you're sure to find at least one mention of the colour stylesheet directive such as `color:#336699;`. The numbers should be interpreted as a triple of values $(33, 66, 99)$, each value describing the amount of red, green, and blue needed to create a given colour. Let us call the colour described by the triple $(33, 66, 99)$ CoolBlue. This convention for colour representation is called the RGB colour model and we can think of it as the *RGB basis*. A basis is a set of elements that can be combined together to express something more complicated. In our case, the **R**, **G**, and **B** elements are pure colours that can create any colour when mixed appropriately. Schematically, we can write this mixing idea as

$$\text{CoolBlue} = (33, 66, 99)_{RGB} = 33\mathbf{R} + 66\mathbf{G} + 99\mathbf{B},$$

where the *components* determine the strength of each colour. To create the colour, we combine its components as symbolized by the $+$ operation.

The cyan, magenta, and yellow (CMY) colour model is another basis for representing colours. To express the "cool blue" colour in the CMY basis, you will need the following components:

$$(33, 66, 99)_{RGB} = \text{CoolBlue} = (222, 189, 156)_{CMY} = 222\mathbf{C} + 189\mathbf{M} + 156\mathbf{Y}.$$

The *same* colour CoolBlue is represented by a *different* set of components when the CMY colour basis is used.

Note that a triple of components by itself doesn't mean anything unless we know the basis being used. For example, if we were to interpret the triple of components $(33, 66, 99)$ with respect to the CMY basis, we would obtain a completely different colour, which would not be cool at all.

A basis is required to convert mathematical objects like the triple (a, b, c) into real-world ideas like colours. As exemplified above, to avoid any ambiguity we can use a subscript after the bracket to indicate the basis associated with each triple of components. Writing $(222, 189, 156)_{CMY}$ and $(33, 66, 99)_{RGB}$ clarifies which basis to use for each triple of components.

Discussion

It would be hard to over-emphasize the importance of the basis—the coordinate system you use to describe vectors. The choice of coordinate system is the bridge between real-world vector quantities and their mathematical representation in terms of components. Every time you start a new problem that involves vector calculations, the first thing you should do is choose the coordinate system you want to use, and indicate it clearly in the diagram.

Using a non-standard coordinate system can sometimes simplify the equations you have to solve. For example, let's say we want to study the motion of a block sliding down an incline with velocity $\vec{v}$, as illustrated in Figure 7.11. Using the standard xy-basis, the velocity vector is represented as $(v\cos\theta, -v\sin\theta)_{xy}$, which has components in both the x- and y-directions and requires using trigonometric functions. If instead you use the non-standard $x'y'$-basis, the components of the velocity will be $(v, 0)_{x'y'}$. Note the velocity only has a component along the x'-direction, which will simplify all subsequent calculations.

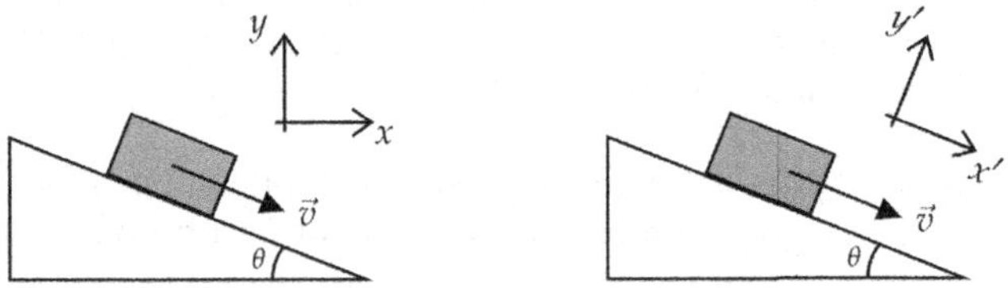

Figure 7.11: The vector $\vec{v}$ is described by the coordinates $(v\cos\theta, -v\sin\theta)_{xy}$ with respect to the standard basis xy. The same vector $\vec{v}$ is described by the coordinates $(v, 0)_{x'y'}$ with respect to the "tilted" basis $x'y'$.

Recall the polar coordinates representation we used to describe points $r\angle\theta$ and vectors $\|\vec{v}\|\angle\theta$ in two dimensions (see page 139). This is another example of an alternative coordinate system that's useful for describing rotations and circular motion. Note certain textbooks will write the polar coordinates of the vector $\vec{v} = \|\vec{v}\|\angle\theta$ using the bracket notation $(\|\vec{v}\|, \theta)$, which can easily be confused with the Cartesian coordinates of the vector (v_x, v_y). Indicating the coordinate system as a subscript after the bracket can avoid any confusion: $\vec{v} = (\|\vec{v}\|, \theta)_{r\theta} = (v_x, v_y)_{xy}$.

Links

[Vectors and vector operations explained by 3Blue1Brown]
`https://www.youtube.com/watch?v=fNk_zzaMoSs`

[More vector illustrations and definitions from Wikipedia]
`https://en.wikipedia.org/wiki/Euclidean_vector`

Exercises

E7.1 Given the vectors $\vec{v}_1 = (2, 1)$, $\vec{v}_2 = (2, -1)$, and $\vec{v}_3 = (3, 3)$, calculate the following expressions:

a) $\vec{v}_1 + \vec{v}_2$ **b)** $\vec{v}_2 - 2\vec{v}_1$ **c)** $\vec{v}_1 + \vec{v}_2 + \vec{v}_3$

E7.2 Express the following vectors as components:

a) $\vec{v}_1 = 10\angle 30°$ **b)** $\vec{v}_2 = 12\angle{-90°}$ **c)** $\vec{v}_3 = 3\angle 170°$

E7.3 Express the following vectors in length-and-direction notation:

a) $\vec{u}_1 = (4, 0)$ **b)** $\vec{u}_2 = (1, 1)$ **c)** $\vec{u}_3 = (-1, 3)$

7.4 Vector products

We'll now define the *dot product* and the *cross product*: two geometric operations useful for working with three-dimensional vectors.

Dot product

The *dot product* takes two vectors as inputs and produces a single, real number as an output:

$$\cdot : \mathbb{R}^3 \times \mathbb{R}^3 \quad \to \quad \mathbb{R}.$$

The dot product between the vector $\vec{v} = (v_x, v_y, v_z)$ and the vector $\vec{w} = (w_x, w_y, w_z)$ can be computed using either the algebraic formula,

$$\vec{v} \cdot \vec{w} \;=\; v_x w_x + v_y w_y + v_z w_z,$$

or the geometric formula,

$$\vec{v} \cdot \vec{w} \;=\; \|\vec{v}\| \|\vec{w}\| \cos(\varphi),$$

where φ is the angle between the two vectors. Note the value of the dot product depends on the vectors' lengths and the cosine of the angle between them.

The name *dot product* comes from the symbol used to denote it. It is also known as the *scalar product*, since the result of the dot product is a scalar number—a number that does not change when the basis changes. The dot product is also sometimes called the *inner product*.

We can combine the algebraic and the geometric formulas for the dot product to obtain the formula,

$$\cos(\varphi) = \frac{\vec{v} \cdot \vec{w}}{\|\vec{v}\| \|\vec{w}\|} = \frac{v_x w_x + v_y w_y + v_z w_z}{\|\vec{v}\| \|\vec{w}\|} \quad \text{and} \quad \varphi = \cos^{-1}(\cos(\varphi)).$$

This formula makes it possible to find the angle between two vectors if we know their components.

The geometric factor $\cos(\varphi)$ depends on the relative orientation of the two vectors as follows:

- If the vectors point in the same direction, then $\cos(\varphi) = \cos(0°) = 1$, so $\vec{v} \cdot \vec{w} = \|\vec{v}\| \|\vec{w}\|$.
- If the vectors are perpendicular to each other, then $\cos(\varphi) = \cos(90°) = 0$, so $\vec{v} \cdot \vec{w} = 0$.
- If the vectors point in exactly opposite directions, then $\cos(\varphi) = \cos(180°) = -1$, so $\vec{v} \cdot \vec{w} = -\|\vec{v}\| \|\vec{w}\|$.

The dot product is defined for vectors of any dimension; as long as two vectors are defined with respect to the same basis, we can compute the dot product between them.

Cross product

The *cross product* takes two vectors as inputs and produces another vector as the output:

$$\times : \mathbb{R}^3 \times \mathbb{R}^3 \quad \to \quad \mathbb{R}^3.$$

The cross product of two vectors is perpendicular to both vectors:

$$\vec{v} \times \vec{w} = \{ \text{ a vector perpendicular to both } \vec{v} \text{ and } \vec{w} \ \} \quad \in \mathbb{R}^3.$$

If you take the cross product of one vector pointing in the x-direction with another vector pointing in the y-direction, the result will be a vector in the z-direction: $\hat{\imath} \times \hat{\jmath} = \hat{k}$. The name *cross product* comes from the symbol used to denote it. It is also sometimes called the *vector product*, since the output of this operation is a vector.

The cross products of individual basis elements are defined as

$$\hat{\imath} \times \hat{\jmath} = \hat{k}, \qquad \hat{\jmath} \times \hat{k} = \hat{\imath}, \qquad \hat{k} \times \hat{\imath} = \hat{\jmath}.$$

Look at Figure 7.10 on page 146 and imagine the vectors $\hat{\imath}$, $\hat{\jmath}$, and $\hat{k}$ pointing along each axis. Try to visualize the three equations above.

The cross product is *anticommutative*, which means swapping the order of the inputs introduces a negative sign in the output:

$$\hat{\jmath} \times \hat{\imath} = -\hat{k}, \qquad \hat{k} \times \hat{\jmath} = -\hat{\imath}, \qquad \hat{\imath} \times \hat{k} = -\hat{\jmath}.$$

It's likely that, until now, the products you've seen in math have been *commutative*, which means the order of the inputs doesn't matter. The product of two numbers is commutative $ab = ba$, and the dot product of two vectors is commutative $\vec{u} \cdot \vec{v} = \vec{v} \cdot \vec{u}$, but the cross product of two vectors is *anti*commutative $\vec{v} \times \vec{w} = -\vec{w} \times \vec{v}$.

Given two vectors $\vec{a} = a_x\hat{\imath} + a_y\hat{\jmath} + a_z\hat{k}$ and $\vec{b} = b_x\hat{\imath} + b_y\hat{\jmath} + b_z\hat{k}$, their cross product is calculated as

$$\vec{a} \times \vec{b} \quad = \quad (a_yb_z - a_zb_y)\hat{\imath} + (a_zb_x - a_xb_z)\hat{\jmath} + (a_xb_y - a_yb_x)\hat{k}.$$

Computing the cross product requires a specific combination of multiplications and subtractions of the input vectors' components. The result of this combination is the vector $\vec{a} \times \vec{b}$ which is perpendicular to both $\vec{a}$ and $\vec{b}$.

The length of the cross product of two vectors is proportional to the sine of the angle between the two vectors:

$$\|\vec{a} \times \vec{b}\| = \|\vec{a}\|\|\vec{b}\| \sin(\varphi).$$

The right-hand rule

Consider the plane formed by the vectors $\vec{a}$ and $\vec{b}$. There are actually *two* vectors perpendicular to this plane: one above the plane and one below the plane. We use the *right-hand rule* to figure out which of these vectors corresponds to the cross product $\vec{a} \times \vec{b}$.

Make a fist with your right hand and then extend your thumb, first finger, and middle finger. When your index finger points in the same direction as the vector $\vec{a}$ and your middle finger points in the direction of $\vec{b}$, your thumb will point in the direction of $\vec{a} \times \vec{b}$. The relationship encoded in the right-hand rule matches the relationship between the standard basis vectors: $\hat{\imath} \times \hat{\jmath} = \hat{k}$.

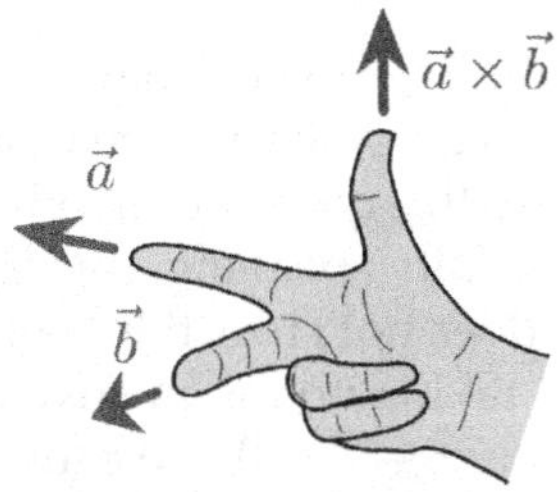

Figure 7.12: Using the right-hand rule to determine the direction of the cross product $\vec{a} \times \vec{b}$ based on directions of $\vec{a}$ and $\vec{b}$.

Links

[Nice illustrations of the cross product]
`http://1ucasvb.tumblr.com/post/76812811092/`
`https://www.youtube.com/watch?v=eu6i7WJeinw`

Exercises

E7.4 Given the vectors $\vec{u} = (1,1,0)$ and $\vec{v} = (0,0,3)$, compute the following vector expressions:

a) $\vec{u} + \vec{v}$ **b)** $\vec{u} - \vec{v}$ **c)** $3\vec{u} + \vec{v}$ **d)** $\|\vec{u}\|$

E7.5 Given $\vec{v} = (1,2,3)$ and $\vec{w} = (0,1,1)$, compute the following vector products: **a)** $\vec{v} \cdot \vec{w}$; **b)** $\vec{v} \times \vec{w}$; **c)** $\vec{w} \times \vec{v}$; **d)** $\vec{w} \times \vec{w}$.

7.5 Complex numbers

By now, you've heard about complex numbers $\mathbb{C}$. The word "complex" is an intimidating word. Surely it must be a complex task to learn about the complex numbers. That may be true in general, but it helps if you know about vectors. Complex numbers are similar to two-dimensional vectors $\vec{v} \in \mathbb{R}^2$. We add and subtract complex numbers like vectors. Complex numbers also have components, length, and "direction." If you understand vectors, you will understand complex numbers at almost no additional mental cost.

We'll begin with a practical problem.

Example

Suppose you're asked to solve the following quadratic equation:

$$x^2 + 1 = 0.$$

You're looking for a number x, such that $x^2 = -1$. If you are only allowed to give real answers (the set of real numbers is denoted $\mathbb{R}$), then there is no answer to this question. In other words, this equation has no solutions. Graphically speaking, this is because the quadratic function $f(x) = x^2 + 1$ does not cross the x-axis.

However, we're not taking no for an answer! If we insist on solving for x in the equation $x^2 + 1 = 0$, we can imagine a new number i that satisfies $i^2 = -1$. We call i the unit imaginary number. The solutions to the equation are therefore $x_1 = i$ and $x_2 = -i$. There are two solutions because the equation is quadratic. We can check that $i^2 + 1 = -1 + 1 = 0$ and also $(-i)^2 + 1 = (-1)^2 i^2 + 1 = i^2 + 1 = 0$.

Thus, while the equation $x^2 + 1 = 0$ has no real solutions, it *does* have solutions if we allow the answers to be imaginary numbers.

Definitions

Complex numbers have a real part and an imaginary part:

- i: the unit imaginary number $i = \sqrt{-1}$ or $i^2 = -1$
- bi: an imaginary number that is equal to b times i
- $\mathbb{R}$: the set of real numbers
- $\mathbb{C}$: the set of complex numbers $\mathbb{C} = \{a + bi \mid a, b \in \mathbb{R}\}$
- $z = a + bi$: a complex number
  - ▷ $\text{Re}\{z\} = a$: the real part of z
  - ▷ $\text{Im}\{z\} = b$: the imaginary part of z
- $\bar{z}$: the *complex conjugate* of z. If $z = a + bi$, then $\bar{z} = a - bi$.

The polar representation of complex numbers:

- $z = |z|\angle\varphi_z = |z|\cos\varphi_z + i|z|\sin\varphi_z$
- $|z| = \sqrt{\bar{z}z} = \sqrt{a^2 + b^2}$: the *magnitude* of $z = a + bi$

- $\varphi_z = \tan^{-1}(b/a)$: the *phase* or *argument* of $z = a + bi$
- $\text{Re}\{z\} = |z| \cos \varphi_z$
- $\text{Im}\{z\} = |z| \sin \varphi_z$

Formulas

Addition and subtraction

Just as we performed the addition of vectors component by component, we perform addition on complex numbers by adding the real parts together and adding the imaginary parts together:

$$(a + bi) + (c + di) = (a + c) + (b + d)i.$$

Polar representation

We can give a geometric interpretation of the complex numbers by extending the real number line into a two-dimensional plane called the *complex plane*. The horizontal axis in the complex plane measures the *real* part of the number. The vertical axis measures the *imaginary* part. Complex numbers are points in the complex plane.

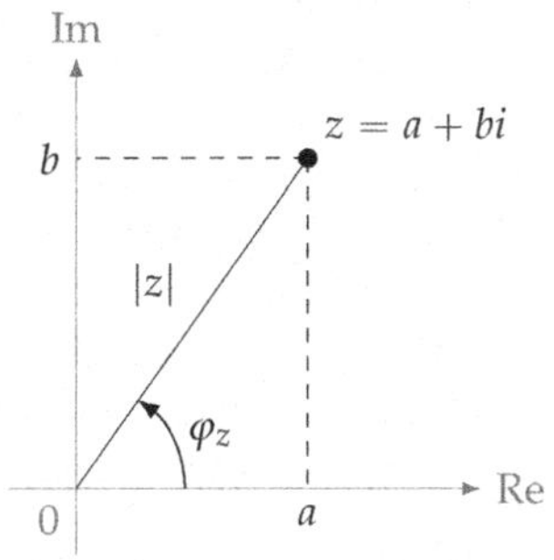

Figure 7.13: The complex number $z = a + bi$ corresponds to the point with coordinates (a, b) in the complex plane.

It is possible to represent any complex number $z = a + bi$ in terms of its *magnitude* and its *phase*:

$$z = |z|\angle\varphi_z = \underbrace{|z|\cos\varphi_z}_{a} + \underbrace{|z|\sin\varphi_z}_{b}\, i.$$

The *magnitude* (or *absolute value*) of a complex number $z = a + bi$ is

$$|z| = \sqrt{a^2 + b^2}.$$

This corresponds to the *length* of the vector that represents the complex number in the complex plane. The formula is obtained by using Pythagoras' theorem.

The *phase*, also known as the *argument* of the complex number $z = a + bi$ is given by the formula

$$\varphi_z = \arg z = \texttt{atan2}(b,a) = \begin{cases} \tan^{-1}\left(\frac{b}{a}\right) & \text{if } a > 0, \\ \pi + \tan^{-1}\left(\frac{b}{a}\right) & \text{if } a < 0, \\ \frac{\pi}{2} & \text{if } a = 0 \text{ and } b > 0, \\ -\frac{\pi}{2} & \text{if } a = 0 \text{ and } b < 0. \end{cases}$$

The phase corresponds to the angle that z forms with the real axis.

We previously saw this complicated-looking formula with four cases when we talked about converting from Cartesian coordinates to polar coordinates for points (Section 6.4) and vectors (Section 7.2). When a certain formula comes up three times in a math book, this should tell you the author *really* wants you to know it. Seriously, do me a favour and revisit the exercise E6.8 (page 115) and the exercise E7.3 (page 150).

Computer algebra systems provide the two-input inverse tangent function `atan2`, which is the easiest way to calculate the phase φ_z for the complex number $z = a + bi$. The function `atan2` handles all four cases automatically and always computes the correct phase φ_z.

* * *

In addition to the vector-like operations we can perform on complex numbers, like computing their magnitude and phase, we can also perform other operations on complex numbers that are not defined for vectors. The set of complex numbers $\mathbb{C}$ is a *field*. This means, in addition to the addition and subtraction operations, we can also perform multiplication and division with complex numbers.

Multiplication

The product of two complex numbers is computed using the usual rules of algebra:

$$\begin{aligned} (a + bi)(c + di) &= a(c + di) + bi(c + di) \\ &= ac + adi + bci + bdi^2 \\ &= (ac - bd) + (ad + bc)i. \end{aligned}$$

In the polar representation, the product formula is

$$(p\angle\phi)(q\angle\psi) = pq\angle(\phi + \psi).$$

To multiply two complex numbers, multiply their magnitudes and add their phases.

Example Verify that $z\bar{z} = a^2 + b^2 = |z|^2$.

Division

Let's look at the procedure for dividing complex numbers:

$$\frac{(a+bi)}{(c+di)} = \frac{(a+bi)}{(c+di)}\frac{(c-di)}{(c-di)} = (a+bi)\frac{(c-di)}{(c^2+d^2)} = (a+bi)\frac{\overline{c+di}}{|c+di|^2}.$$

In other words, to divide the number z by the complex number s, compute $\bar{s}$ and $|s|^2 = s\bar{s}$ and then use

$$z/s = z\frac{\bar{s}}{|s|^2}.$$

You can think of $\frac{\bar{s}}{|s|^2}$ as being equivalent to s^{-1}.

Cardano's example One of the earliest examples of reasoning involving complex numbers was given by Gerolamo Cardano in his 1545 book *Ars Magna*. Cardano wrote, "If someone says to you, divide 10 into two parts, one of which multiplied into the other shall produce 40, it is evident that this case or question is impossible." We want to find numbers x_1 and x_2 such that $x_1 + x_2 = 10$ and $x_1x_2 = 40$. This sounds kind of impossible. Or is it?

"Nevertheless," Cardano said, "we shall solve it in this fashion:

$$x_1 = 5 + \sqrt{15}i \text{ and } x_2 = 5 - \sqrt{15}i."$$

When you add $x_1 + x_2$ you obtain 10. When you multiply the two numbers the answer is

$$\begin{aligned} x_1x_2 &= \left(5+\sqrt{15}i\right)\left(5-\sqrt{15}i\right) \\ &= 25 - 5\sqrt{15}i + 5\sqrt{15}i - \sqrt{15}^2 i^2 = 25 + 15 = 40. \end{aligned}$$

Hence $5 + \sqrt{15}i$ and $5 - \sqrt{15}i$ are two numbers whose sum is 10 and whose product is 40.

Example 2 Let's compute the product of -1 and i. The answer is obviously $-i$, but let's look at this simple calculation geometrically. The polar representation of the number i is $1\angle\frac{\pi}{2}$. Multiplication of any complex number $z = |z|\angle\varphi_z$ by i corresponds to adding $\frac{\pi}{2}$ to the phase of the number:

$$zi = (|z|\angle\varphi_z)(1\angle\tfrac{\pi}{2}) = (|z|\cdot 1)\angle(\varphi_z + \tfrac{\pi}{2}) = |z|\angle(\varphi_z + \tfrac{\pi}{2}).$$

In other words, multiplication by i is equivalent to applying a $\frac{\pi}{2}$ (90°) counterclockwise rotation in the complex plane. We can therefore interpret the answer $(-1)(i) = -i$ as the complex number $-1 = 1\angle\pi$ experiencing a $\frac{\pi}{2}$ rotation to arrive at $1\angle(\pi + \frac{\pi}{2}) = 1\angle\frac{3\pi}{2} = -i$.

Example 3 Find the polar representation of $z = -3 - i$ and compute z^6. Let's denote the polar representation of z by $z = r\angle\varphi$ as shown in Figure 7.14. We find $r = \sqrt{3^2 + 1^2} = \sqrt{10}$ and $\varphi = \tan^{-1}(\frac{1}{3}) + \pi = 0.322 + \pi$. Using the polar representation, we can easily compute z^6:

$$z^6 = r^6\angle(6\varphi) = (\sqrt{10})^6\angle 6(0.322 + \pi) = 10^3\angle 1.932 + 6\pi = 10^3\angle 1.932.$$

Note we can ignore multiples of 2π in the phase. We thus find the value of z^6 is $1000\cos(1.932) + 1000\sin(1.932)i = -353.4 + 935.5i$.

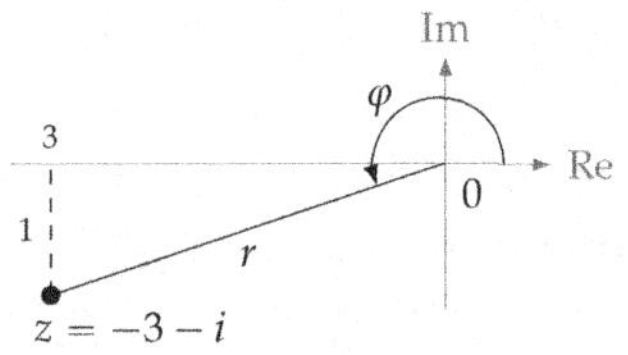

Figure 7.14: The complex number $z = 3 - i$ has magnitude $r = \sqrt{10}$ and phase $\varphi = 0.322 + \pi = 3.463$.

Fundamental theorem of algebra

The fundamental theorem of algebra states that any polynomial of degree n, $P(x) = a_n x^n + \cdots + a_2 x^2 + a_1 x + a_0$, can be written as

$$P(x) = a_n(x - z_1)(x - z_2)\cdots(x - z_n),$$

where $z_i \in \mathbb{C}$ are the polynomial's roots. In other words, the equation $P(x) = 0$ has n solutions: the complex numbers $z_1, z_2, \ldots, z_n$. Before today, you might have said the equation $x^2 + 1 = 0$ has no solutions. Now you know its solutions are the complex numbers $z_1 = i$ and $z_2 = -i$.

The theorem is "fundamental" because it tells us we'll never need to invent numbers "fancier" than the complex numbers to solve polynomial equations. To understand why this is important, recall that each set of numbers is associated with a different class of equations. Figure 1.1 on page 8 shows the nested containment structure of the number sets $\mathbb{N}$, $\mathbb{Z}$, $\mathbb{Q}$, $\mathbb{R}$, and $\mathbb{C}$. The natural numbers $\mathbb{N}$ appear as solutions of the equation $m + n = x$, where m and n are natural numbers (denoted $m, n \in \mathbb{N}$). The integers $\mathbb{Z}$ are the solutions to equations of the form $x + m = n$, where $m, n \in \mathbb{N}$. The rational numbers $\mathbb{Q}$ are necessary to solve for x in $mx = n$, with $m, n \in \mathbb{Z}$. To find the solutions of $x^2 = 2$, we need the real numbers $\mathbb{R}$. And in this section, we learned that the solutions to the equation $x^2 = -1$ are complex numbers $\mathbb{C}$. At this point you might be wondering if you're attending some sort of math party, where mathematicians write down complicated equations and—just for the fun of it—invent new sets of numbers to describe the solutions to these equations. Can this process continue indefinitely?

Nope. The party ends with $\mathbb{C}$. The fundamental theorem of algebra guarantees that any polynomial equation you could come up with—no matter how complicated it is—has solutions that are complex numbers $\mathbb{C}$.

Euler's formula

It turns out the exponential function is related to the functions sine and cosine. Lo and behold, we have *Euler's formula*:

$$e^{i\theta} = \cos\theta + i\sin\theta\,.$$

Inputting an imaginary number to the exponential function outputs a complex number that contains both cos and sin. Euler's formula gives us an alternate notation for the polar representation of complex numbers: $z = |z|\angle\varphi_z = |z|e^{i\varphi_z}$.

If you want to impress your friends with your math knowledge, plug $\theta = \pi$ into the above equation to find

$$e^{i\pi} = \cos(\pi) + i\sin(\pi) = -1,$$

which can be rearranged to obtain the equation $e^{i\pi} + 1 = 0$. The equation $e^{i\pi} + 1 = 0$ is called *Euler's identity*, and it shows a relationship between the five most important numbers in all of mathematics: Euler's number $e = 2.71828\ldots$, $\pi = 3.14159\ldots$, the imaginary number i, 1, and zero. It's kind of cool to see all these important numbers reunited in one equation, don't you agree?

One way to understand the equation $e^{i\pi} + 1 = 0$, is to think of $e^{i\pi}$ as the polar representation of the complex number $z = 1e^{i\pi} = 1\angle\pi$, which is the same as 1 rotated counterclockwise by π radians (180°) in the complex plane. We know $e^{i\pi} = 1\angle\pi = -1$ and so $e^{i\pi} + 1 = 0$.

De Moivre's formula

By replacing θ in Euler's formula with $n\theta$, we obtain de Moivre's formula:

$$(\cos\theta + i\sin\theta)^n = \cos n\theta + i\sin n\theta.$$

De Moivre's formula makes sense if you think of the complex number $z = e^{i\theta} = \cos\theta + i\sin\theta$, raised to the nth power:

$$(\cos\theta + i\sin\theta)^n = z^n = (e^{i\theta})^n = e^{in\theta} = \cos n\theta + i\sin n\theta.$$

Setting $n = 2$ in de Moivre's formula, we can derive the double angle formulas (page 106) as the real and imaginary parts of the following equation:

$$(\cos^2\theta - \sin^2\theta) + (2\sin\theta\cos\theta)i = \cos(2\theta) + \sin(2\theta)i.$$

Links

[Intuitive proof of the fundamental theorem of algebra]
`https://www.youtube.com/watch?v=shEk8sz1oOw`

Chapter 8

Extra topics

Learning all of mathematics would require several lifetimes. There is so much to learn: applied math, abstract theoretical math, numerical methods, and many other subfields and specialties. It's up to you to choose how much you want to learn. As an adult learner there will be no exams to force you to study, so if you're learning something it's because you *want* to know!

Out of the thousands of possible math topics you could learn next, I've selected a shortlist of three important topics to get you started. In Section 8.1 we'll explore how to solve equations with multiple unknowns. Specifically, we'll focus on systems of k linear equations that contain k unknowns. For example, the *system of equations*

$$\begin{aligned} 1s + 2t &= 5 \\ 3s + 9t &= 21 \end{aligned}$$

consists of two equations that contain two unknowns: s and t. There is a systematic procedure you can follow to combine the equations and reduce the problem to a single equation with a single unknown

$$3t = 6,$$

which you know how to solve. In the end, we find the unknowns s and t that satisfy both equations are $s = 1$ and $t = 2$.

In Section 8.2, we'll learn about the compound interest calculations used by banks to compute the interest owed on loans. It's important to understand the math so you can calculate the cost of borrowing for different interest rates and compounding methods. This topic is particularly relevant for students who are considering taking out student loans. Think of this math knowledge as financial self-defence.

Finally, in Section 8.3, we'll introduce sets and set notation. Mathematicians often use symbols like $\in$ (element of), $\subset$ (subset of), $\forall$ (for all), and $\exists$ (there exists), to make very concise math statements and definitions. We managed to get through the entire book without the need for these "alien symbols," but knowing set notation is important for your future math studies. Most advanced math textbooks assume readers are familiar with these symbols, so it's a good idea to know what the symbols mean.

8.1 Solving systems of linear equations

Solving equations with one unknown—like $2x + 4 = 7x$, for instance—requires manipulating both sides of the equation until the unknown variable is *isolated* on one side. For this instance, we can subtract $2x$ from both sides of the equation to obtain $4 = 5x$, which simplifies to $x = \frac{4}{5}$.

What about the case when you are given *two* equations and must solve for *two* unknowns? For example,

$$x + 2y = 5,$$
$$3x + 9y = 21.$$

Can you find values of x and y that satisfy both equations?

Concepts

- x, y: the two unknowns in the equations
- $eq1, eq2$: a system of two equations that must be solved *simultaneously*. These equations will look like

$$a_1x + b_1y = c_1,$$
$$a_2x + b_2y = c_2,$$

where as, bs, and cs are given constants.

Principles

If you have n equations and n unknowns, you can solve the equations *simultaneously* and find the values of the unknowns. There are several different approaches for solving equations simultaneously. We'll show three of these approaches for the case $n = 2$.

Solution techniques

When solving for two unknowns in two equations, the best approach is to *eliminate* one of the variables from the equations. By combining the two equations appropriately, we can simplify the problem to the problem of finding one unknown in one equation.

Solving by substitution

We want to solve the following system of equations:

$$\begin{aligned} x + 2y &= 5, \\ 3x + 9y &= 21. \end{aligned}$$

We can isolate x in the first equation to obtain

$$\begin{aligned} x &= 5 - 2y, \\ 3x + 9y &= 21. \end{aligned}$$

Now *substitute* the expression for x from the top equation into the bottom equation:

$$3(5 - 2y) + 9y = 21.$$

We just eliminated one of the unknowns by substitution. Continuing, we expand the bracket to find

$$15 - 6y + 9y = 21,$$

or

$$3y = 6.$$

We find $y = 2$, but what is x? Easy. To solve for x, plug the value $y = 2$ into any of the equations we started from. Using the equation $x = 5 - 2y$, we find $x = 5 - 2(2) = 1$.

Solving by subtraction

Let's now look at another way to solve the same system of equations:

$$\begin{aligned} x + 2y &= 5, \\ 3x + 9y &= 21. \end{aligned}$$

Observe that any equation will remain true if we multiply the whole equation by some constant. For example, we can multiply the first equation by 3 to obtain an equivalent set of equations:

$$\begin{aligned} 3x + 6y &= 15, \\ 3x + 9y &= 21. \end{aligned}$$

Why did I pick 3 as the multiplier? By choosing this constant, the x terms in both equations now have the same coefficient.

Subtracting two true equations yields another true equation. Let's subtract the top equation from the bottom one:

$$\cancel{3x} - \cancel{3x} + 9y - 6y = 21 - 15 \quad \Rightarrow \quad 3y = 6.$$

The $3x$ terms cancel. This subtraction eliminates the variable x because we multiplied the first equation by 3. We find $y = 2$. To find x, substitute $y = 2$ into one of the original equations:

$$x + 2(2) = 5,$$

from which we deduce that $x = 1$.

Solving by equating

There is a third way to solve the system of equations

$$\begin{aligned} x + 2y &= 5, \\ 3x + 9y &= 21. \end{aligned}$$

We can isolate x in both equations by moving all other variables and constants to the right-hand sides of the equations:

$$\begin{aligned} x &= 5 - 2y, \\ x &= \frac{1}{3}(21 - 9y) = 7 - 3y. \end{aligned}$$

Though the variable x is unknown to us, we know two facts about it: x is equal to $5 - 2y$ and x is equal to $7 - 3y$. Therefore, we can eliminate x by equating the right-hand sides of the equations:

$$5 - 2y = 7 - 3y.$$

We solve for y by adding $3y$ to both sides and subtracting 5 from both sides. We find $y = 2$ then plug this value into the equation $x = 5 - 2y$ to find x. The solutions are $x = 1$ and $y = 2$.

Discussion

The repeated use of the three algebraic techniques presented in this section will allow you to solve any system of n linear equations in n unknowns. Each time you eliminate one variable using a substitution, a subtraction, or an elimination by equating, you're simplifying the problem to a problem of finding $(n - 1)$ unknowns in a system of $(n - 1)$ equations. There is an entire math course called linear algebra, in which you'll develop a systematic approach for solving systems of linear equations.

Geometric solution

Solving a system of two linear equations in two unknowns can be understood geometrically as finding the point of intersection between two lines in the Cartesian plane. In this section we'll explore this correspondence between algebra and geometry to develop yet another way of solving systems of linear equations.

The algebraic equation $ax + by = c$ containing the unknowns x and y can be interpreted as a *constraint* equation on the set of possible values for the variables x and y. We can visualize this constraint geometrically by considering the coordinate pairs (x, y) that lie in the Cartesian plane. Recall that every point in the Cartesian plane can be represented as a coordinate pair (x, y), where x and y are the coordinates of the point.

Figure 8.1 shows the geometrical representation of three equations. The line ℓ_a corresponds to the set of points (x, y) that satisfy the equation $x = 1$, the line ℓ_b is the set of points (x, y) that satisfy the equation $y = 2$, and the line ℓ_c corresponds to the set of points that satisfy $x + 2y = 2$.

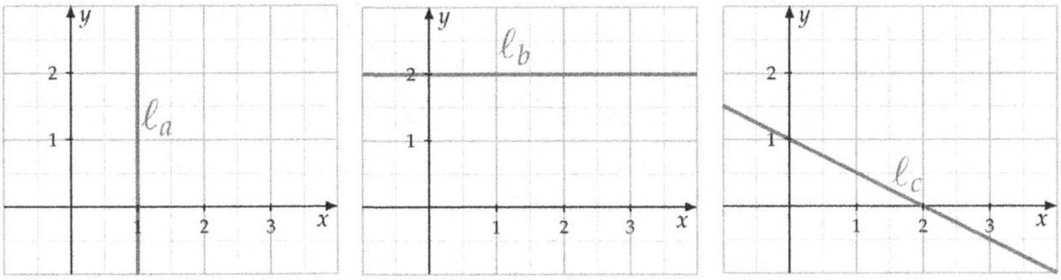

(a) $x = 1$ (b) $y = 2$ (c) $x + 2y = 2$

Figure 8.1: Graphical representations of three linear equations.

You can convince yourself that the geometric lines shown in Figure 8.1 are equivalent to the algebraic equations by considering individual points (x, y) in the plane. For example, the points $(1, 0)$, $(1, 1)$, and $(1, 2)$ are all part of the line ℓ_a since they satisfy the equation $x = 1$. For the line ℓ_c, you can verify that the line's x-intercept $(2, 0)$ and its y-intercept $(0, 1)$ both satisfy the equation $x + 2y = 2$.

The Cartesian plane as a whole corresponds to the set $\mathbb{R}^2$, which describes all possible pairs of coordinates. To understand the equivalence between the algebraic equation $ax + by = c$ and the line ℓ in the Cartesian plane, we can use the following precise math notation:

$$\ell : \{(x, y) \in \mathbb{R}^2 \mid ax + by = c\}.$$

In words, this means that the line ℓ is defined as the subset of the pairs of real numbers (x, y) that satisfy the equation $ax + by = c$.

Figure 8.2 shows the graphical representation of the line ℓ.

You don't have to take my word for it, though! Think about it and convince yourself that all points on the line ℓ shown in Figure 8.2 satisfy the equation $ax + by = c$. For example, you can check that the x-intercept $(\frac{c}{a}, 0)$ and the y-intercept $(0, \frac{c}{b})$ satisfy the equation $ax + by = c$.

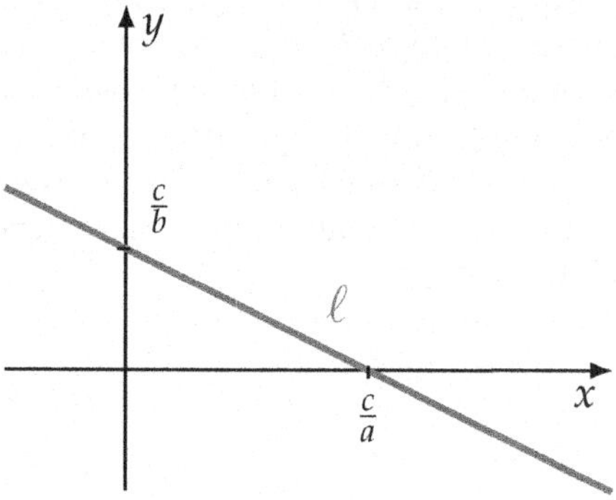

Figure 8.2: Graphical representation of the equation $ax + by = c$.

Solving the system of two equations

$$a_1x + b_1y = c_1,$$
$$a_2x + b_2y = c_2,$$

corresponds to finding the intersection of the lines ℓ_1 and ℓ_2 that represent each equation. The pair (x, y) that satisfies both algebraic equations simultaneously is equivalent to the point (x, y) that is the intersection of lines ℓ_1 and ℓ_2, as illustrated in Figure 8.3.

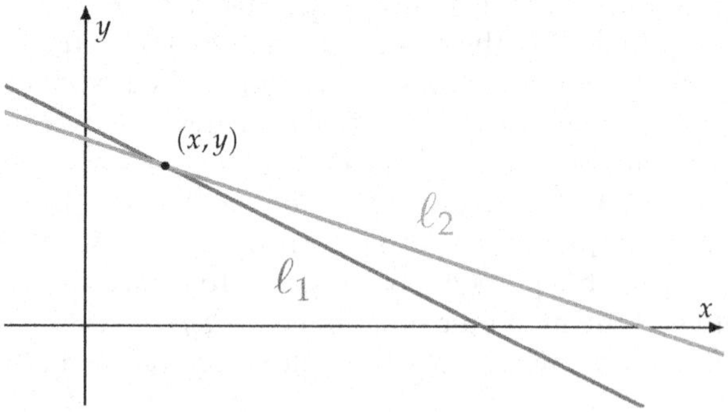

Figure 8.3: The point (x, y) that lies at the intersection of lines ℓ_1 and ℓ_2.

Example Let's see how we can use the geometric interpretation to solve the system of equations

$$x + 2y = 5,$$
$$3x + 9y = 21.$$

We've already seen three different *algebraic* techniques for finding the solution to this system of equations; now let's see a *geometric* approach for finding the solution. I'm not kidding you, we're going to solve the exact same system of equations a fourth time!

The first step is to draw the lines that correspond to each of the equations using pen and paper or a graphing calculator. The second step is to find the coordinates of the point where the two lines intersect as shown in Figure 8.4. The point $(1,2)$ that lies on both lines ℓ_1 and ℓ_2 corresponds to the x and y values that satisfy both equations simultaneously.

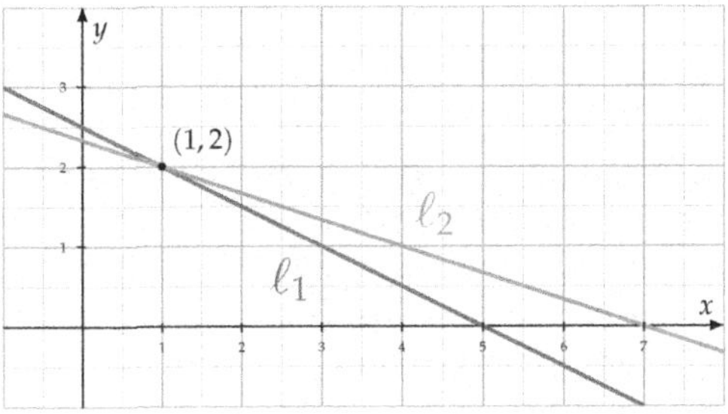

Figure 8.4: The line ℓ_1 with equations $x + 2y = 5$ intersects the line ℓ_2 with equation $3x + 9y = 21$ at the point $(1,2)$.

Visit the webpage at `www.desmos.com/calculator/exikik615f` to play with an interactive version of the graphs shown in Figure 8.4. Try changing the equations and see how the graphs change.

Exercises

E8.1 Plot the lines ℓ_a, ℓ_b, and ℓ_c shown in Figure 8.1 (page 165) using the Desmos graphing calculator. Use the graphical representation of these lines to find: **a)** the intersection of lines ℓ_c and ℓ_a, **b)** the intersection of ℓ_a and ℓ_b, and **c)** the intersection of lines ℓ_b and ℓ_c.

E8.2 Solve the system of equations simultaneously for x and y:

$$2x + 4y = 16,$$
$$5x - y = 7.$$

E8.3 Solve the system of equations for the unknowns x, y, and z:

$$\begin{aligned} 2x + y - 4z &= 28, \\ x + y + z &= 8, \\ 2x - y - 6z &= 22. \end{aligned}$$

E8.4 Solve for p and q given the equations $p + q = 10$ and $p - q = 4$.

8.2 Compound interest

Soon after ancient civilizations invented the notion of numbers, they started computing *interest* on loans. It is a good idea to know how interest calculations work so that you will be able to make informed decisions about your finances.

Percentages

We often talk about ratios between quantities, rather than mentioning the quantities themselves. For example, we can imagine average Joe, who invests \$1000 in the stock market and loses \$300 because the boys on Wall Street keep pulling dirty tricks on him. To put the number \$300 into perspective, we can say Joe lost 0.3 of his investment, or alternatively, we can say Joe lost 30% of his investment.

To express a ratio as a percentage, multiply it by 100. The ratio of Joe's loss to investment is

$$R = 300/1000 = 0.3.$$

The same ratio expressed as a percentage gives

$$R = 300/1000 \times 100 = 30\%.$$

To convert from a percentage to a ratio, divide the percentage by 100.

Interest rates

Say you take out a \$1000 loan with an interest rate of 6% compounded annually. How much will you owe in interest at the end of the year?

Since 6% corresponds to a ratio of 6/100, and since you borrowed \$1000, the accumulated interest at the end of the year will be

$$I_1 = \frac{6}{100} \times \$1000 = \$60.$$

At year's end, you'll owe the bank a total of

$$L_1 = \left(1 + \frac{6}{100}\right) 1000 = (1 + 0.06)1000 = 1.06 \times 1000 = \$1060.$$

The total money owed after 6 years will be

$$L_6 = (1.06)^6 \times 1000 = \$1418.52.$$

You borrowed \$1000, but in six years you will need to give back \$1418.52. This is a terrible deal! But it gets worse. The above scenario assumes that the bank compounds interest only once per year. In practice, interest is compounded each month.

Monthly compounding

An annual compounding schedule is disadvantageous to the bank, and since the bank writes the rules, compounding is usually performed every month.

The monthly interest rate can be used to find the annual rate. The bank quotes the *nominal annual percentage rate* (APR), which is equal to

$$\text{nominal APR} = 12 \times r,$$

where r is the monthly interest rate.

Suppose we have a nominal APR of 6%, which gives a monthly interest rate of $r = 0.5\%$. If you borrow \$1000 at that interest rate, at the end of the first year you will owe

$$L_1 = \left(1 + \frac{0.5}{100}\right)^{12} \times 1000 = \$1061.68,$$

and after 6 years you will owe

$$L_6 = \left(1 + \frac{0.5}{100}\right)^{72} \times 1000 = 1.061677^6 \times 1000 = \$1432.04.$$

Note how the bank tries to pull a fast one: the *effective* APR is actually 6.16%, not 6%. Every twelve months, the amount due will increase by the following factor:

$$\text{effective APR} = \left(1 + \frac{0.5}{100}\right)^{12} = 1.0616.$$

Thus the effective annual percent rate is 6.16%, but it's legal for banks to advertise it as "6% nominal APR." Sneaky stuff.

Compounding infinitely often

We saw that more frequent compounding leads to higher effective interest rates. Let's find a formula for the effective APR if the nominal APR is 6% and the bank performs the compounding n times per year.

The annual growth ratio will be

$$\left(1+\frac{6}{100n}\right)^n,$$

where the interest rate per compounding period is $\frac{6}{n}$%, and there are n periods per year.

Consider a scenario in which the compounding is performed infinitely often. This corresponds to the case when the number n in the above equation tends to infinity (denoted $n \to \infty$). This scenario leads to the definition of the exponential function $f(x) = e^x$.

When we set $n \to \infty$ in the above expression, the annual growth ratio is described by the exponential function base e as follows:

$$\lim_{n\to\infty}\left(1+\frac{6}{100n}\right)^n = \exp\!\left(\frac{6}{100}\right) = 1.0618365.$$

The expression "$\lim_{n\to\infty}$" is to be read as "In the limit when n tends to infinity."

A nominal APR of 6% with compounding that occurs infinitely often has effective APR of 6.183%. After six years you will owe

$$L_6 = \exp\!\left(\frac{6}{100}\right)^6 \times 1000 = \$1433.33.$$

The nominal APR is 6% in each case, yet, the more frequent the compounding schedule, the more money you'll owe after six years.

Exercises

E8.5 Studious Jack borrowed \$40 000 to complete his university studies and made no payments since graduation. Calculate how much money he owes after 10 years in each of the scenarios.

a) Nominal annual interest rate of 3% compounded monthly

b) Effective annual interest rate of 4%

c) Nominal annual interest rate of 5% with infinite compounding

E8.6 Entrepreneurial Kate borrowed \$20 000 to start a business. Initially her loan had an effective annual percentage rate of 6%, but after five years she negotiated with the bank to obtain a lower rate of 4%. How much money does she owe after 10 years?

8.3 Set notation

A *set* is the mathematically precise notion for describing a group of objects. You don't need to know about sets to perform simple math; but more advanced topics require an understanding of what sets are and how to denote set membership, set operations, and set containment relations. This section introduces all the relevant concepts.

Definitions

- *set*: a collection of mathematical objects
- S, T: the usual variable names for sets
- $s \in S$: this statement is read "s is an element of S" or "s is in S"
- $\mathbb{N}, \mathbb{Z}, \mathbb{Q}, \mathbb{R}$: some important number sets: the naturals, the integers, the rationals, and the real numbers, respectively.
- $\varnothing$: the *empty set* is a set that contains no elements
- $\{ \ldots\ldots \}$: the curly brackets are used to define sets, and the expression inside the curly brackets describes the set contents.

Set operations:

- $S \cup T$: the *union* of two sets. The union of S and T corresponds to the elements in either S or T.
- $S \cap T$: the *intersection* of the two sets. The intersection of S and T corresponds to the elements that are in both S and T.
- $S \setminus T$: *set difference* or *set minus*. The set difference $S \setminus T$ corresponds to the elements of S that are not in T.

Set relations:

- $\subset$: is a strict subset of
- $\subseteq$: is a subset of or equal to

Here is a list of special mathematical shorthand symbols and their corresponding meanings:

- $\in$: element of
- $\notin$: not an element of
- $\forall$: for all
- $\exists$: there exists
- $\nexists$: there doesn't exist
- $|$: such that

These symbols are used in math proofs because they allow us to express complex mathematical arguments succinctly and precisely.

An *interval* is a subset of the real line. We denote an interval by specifying its endpoints and surrounding them with either square brackets "[" or round brackets "(" to indicate whether or not the corresponding endpoint is included in the interval.

- $[a, b]$: the *closed* interval from a to b. This corresponds to the set of numbers between a and b on the real line, including the endpoints a and b. $[a, b] = \{x \in \mathbb{R} \mid a \leqslant x \leqslant b\}$.
- (a, b): the *open* interval from a to b. This corresponds to the set of numbers between a and b on the real line, *not* including the endpoints a and b. $(a, b) = \{x \in \mathbb{R} \mid a < x < b\}$.
- $[a, b)$: the half-open interval that includes the left endpoint a but not the right endpoint b. $[a, b) = \{x \in \mathbb{R} \mid a \leqslant x < b\}$.

Sometimes we encounter intervals that consist of two disjointed parts. We use the notation $[a, b] \cup [c, d]$ to denote the union of the two intervals, which is the set of numbers *either* between a and b (inclusive) *or* between c and d (inclusive).

Sets

Much of math's power comes from *abstraction*: the ability to see the bigger picture and think *meta* thoughts about the common relationships between math objects. We can think of individual numbers like 3, -5, and π, or we can talk about the set of *all* numbers.

It is often useful to restrict our attention to a specific *subset* of the numbers as in the following examples.

Example 1: The nonnegative real numbers

Define $\mathbb{R}_+ \subset \mathbb{R}$ (read "$\mathbb{R}_+$ is a subset of $\mathbb{R}$") to be the set of nonnegative real numbers:

$$\mathbb{R}_+ \stackrel{\text{def}}{=} \{\text{all } x \text{ in } \mathbb{R} \text{ such that } x \geqslant 0\},$$

or expressed more compactly,

$$\mathbb{R}_+ \stackrel{\text{def}}{=} \{x \in \mathbb{R} \mid x \geqslant 0\}.$$

If we were to translate the above expression into plain English, it would read "The set $\mathbb{R}_+$ is defined as the set of all real numbers x such that x is greater or equal to zero."

Note we used the "is defined as" symbol "$\stackrel{\text{def}}{=}$" instead of the basic "$=$" to give an extra hint that we're defining a new variable $\mathbb{R}_+$ that is equal to the set expression on the right. In this book, we'll sometimes use the symbol "$\stackrel{\text{def}}{=}$" when defining new variables and math quantities. Some other books use the notation "$:=$" or "$\equiv$" for this purpose. The meaning of "$\stackrel{\text{def}}{=}$" is identical to "$=$" but it tells us the variable on the left of the equality is new.

Example 2: Even and odd integers

Define the set of even integers as

$$E \stackrel{\text{def}}{=} \{m \in \mathbb{Z} \mid m = 2n,\ n \in \mathbb{Z}\} = \{\ldots, -4, -2, 0, 2, 4, \ldots\}$$

and the set of odd integers as

$$O \stackrel{\text{def}}{=} \{m \in \mathbb{Z} \mid m = 2n + 1,\ n \in \mathbb{Z}\} = \{\ldots, -3, -1, 1, 3, 5, \ldots\}.$$

Indeed, every even number is divisible by two, so it can be written in the form $2n$ for some integer n. Odd numbers can be obtained from the "template" $2n + 1$, with n varying over all integers.

In both of the above examples, we use the *set-builder* notation $\{\ldots \mid \ldots\}$ to define the sets. Inside the curly braces we first describe the general kind of mathematical objects we are talking about, followed by the symbol "$|$" (read "such that"), followed by the conditions that must be satisfied by all elements of the set.

Number sets

Recall the fundamental number sets we defined in Section 1.2 in the beginning of the book. It is worthwhile to review them briefly.

The *natural* numbers form the set derived when you start from 0 and add 1 any number of times:

$$\mathbb{N} \stackrel{\text{def}}{=} \{0, 1, 2, 3, 4, 5, 6, \ldots\}.$$

We use the notation $\mathbb{N}^*$ to denote the set of *positive natural numbers*. The set $\mathbb{N}^*$ is the same as $\mathbb{N}$ but excludes zero.

The integers are the numbers derived by adding or subtracting 1 some number of times:

$$\mathbb{Z} \stackrel{\text{def}}{=} \{x \mid x = \pm n, n \in \mathbb{N}\}.$$

If we allow for divisions between integers, we require the set of rational numbers to represent the results:

$$\mathbb{Q} \stackrel{\text{def}}{=} \left\{ \frac{m}{n} \,\middle|\, m \in \mathbb{Z},\ n \in \mathbb{N}^* \right\},$$

In words, this expression is telling us that every rational number can be written as a fraction $\frac{m}{n}$, where m is an integer ($m \in \mathbb{Z}$), and n is a positive natural number ($n \in \mathbb{N}^*$).

The broader class of real numbers also includes all rationals as well as irrational numbers like $\sqrt{2}$ and π:

$$\mathbb{R} \stackrel{\text{def}}{=} \{\pi, e, -1.53929411\ldots,\ 4.99401940129401\ldots,\ \ldots\}.$$

Finally, we have the set of complex numbers:

$$\mathbb{C} \stackrel{\text{def}}{=} \{1, i, 1+i, 2+3i, \ldots\},$$

where $i \stackrel{\text{def}}{=} \sqrt{-1}$ is the unit imaginary number.

Note that the definitions of $\mathbb{R}$ and $\mathbb{C}$ are not very precise. Rather than give a precise definition of each set inside the curly braces as we did for $\mathbb{Z}$ and $\mathbb{Q}$, we instead stated some examples of the elements in the set. Mathematicians sometimes do this and expect you to guess the general pattern for all the elements in the set.

The following inclusion relationship holds for the fundamental sets of numbers:

$$\mathbb{N} \subset \mathbb{Z} \subset \mathbb{Q} \subset \mathbb{R} \subset \mathbb{C}.$$

This relationship means every natural number is also an integer. Every integer is a rational number. Every rational number is a real. And every real number is also a complex number. See Figure 1.1 (page 8) for an illustration of the subset relationship between the number sets.

Rational numbers and fractions

So far in this book, we've used the notions of "fraction" and "rational number" somewhat interchangeably. Now that we've learned about sets, we can clarify the differences and equivalencies between these related concepts.

The same rational number $\frac{2}{3}$ can be written as a fraction in multiple, equivalent ways. The fractions $\frac{2}{3}$, $\frac{4}{6}$, $\frac{6}{9}$, $\frac{8}{12}$, and $\frac{2k}{3k}$ all correspond to the same rational number. Keep in mind the existence of these *equivalent fractions* whenever checking whether two rational numbers are equal. For example, one person could obtain the answer $\frac{2}{3}$ to a given problem, while another person obtains the answer $\frac{4}{6}$. Since the two fractions look different, we might think these are different answers, when in fact both answers correspond to the same rational number.

A *reduced fraction* is a fraction of the form $\frac{m}{n}$ such that the numbers m and n are the smallest possible. We can obtain the reduced fraction by getting rid of any common factors that appear both in the numerator and denominator. For example,

$$\frac{4}{6} = \frac{2 \cdot 2}{3 \cdot 2} = \frac{2 \cdot \cancel{2}}{3 \cdot \cancel{2}} = \frac{2}{3},$$

where we cancelled the common factor 2 to obtain the equivalent reduced fraction. Reduced fractions are a useful representation for the set of rational numbers, because each rational number corresponds to a unique reduced fraction. Two rational numbers are equal if and only if they correspond to the same reduced fraction.

Subsets of the real line

Recall that the real numbers $\mathbb{R}$ have a graphical representation as points on the number line. See Figure 1.12 on page 22 for a reminder. The number line is also useful for representing various subsets of the real numbers, which we call *intervals*. We can graphically represent an interval by setting a section of the number line in **bold**. For example, the set of numbers that are strictly greater than 2 and strictly smaller than 4 is represented mathematically either as "$(2, 4)$," or more explicitly as

$$\{x \in \mathbb{R} \mid 2 < x < 4\},$$

or graphically as in Figure 8.5.

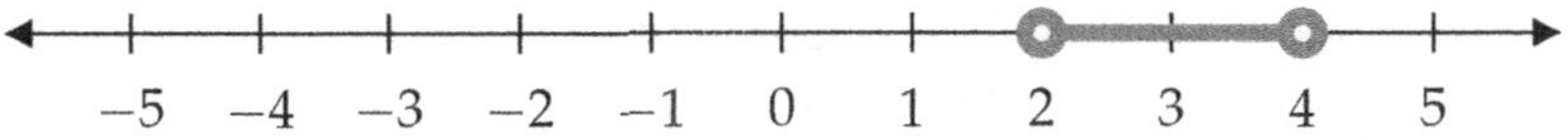

Figure 8.5: The open interval $(2, 4) = \{x \in \mathbb{R} \mid 2 < x < 4\}$.

Let's read the mathematical definition of this set carefully, and try to connect it with the graphical representation. Recall that the symbol $\in$ denotes set membership and the vertical bar stands for "such that," so the whole expression "$\{x \in \mathbb{R} \mid 2 < x < 4\}$" is read "the set of real numbers x, such that $2 < x < 4$." Indeed this is also the region shown in bold in Figure 8.5.

Note that this interval is described by *strict* inequalities, which means the subset contains 2.000000001 and 3.99999999, but doesn't contain the endpoints 2 and 4. These *open* endpoints 2 and 4 are denoted on the number line as empty dots. An empty dot indicates that the endpoint is not included in the set.

We use the *union* symbol ($\cup$) to denote subsets of the number line that consist of several parts. For example, the set of numbers that lies *either* between -3 and 0 *or* between 1 and 2 is written as

$$\{x \in \mathbb{R} \mid -3 \leqslant x \leqslant 0\} \cup \{x \in \mathbb{R} \mid 1 \leqslant x \leqslant 2\}.$$

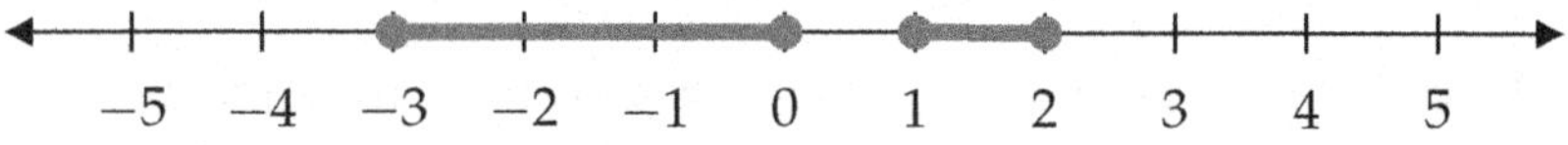

Figure 8.6: The graphical representation of the set $[-3,0] \cup [1,2]$.

This set is defined by less-than-or-equal inequalities, so the intervals contain their endpoints. These *closed* endpoints are denoted on the number line with filled-in dots.

Set relations

We'll now introduce a useful graphical representation for set relations and set operations. Although sets are purely mathematical constructs and they have no "shape," we can draw *Venn diagrams* to visualize relationships between sets and different subsets.

Consider the notion of a set B that is strictly contained in another set A. We write $B \subset A$ if $\forall b \in B$, $b \in A$ as well. Written in words, $B \subset A$ tells us every element of B is also an element of A.

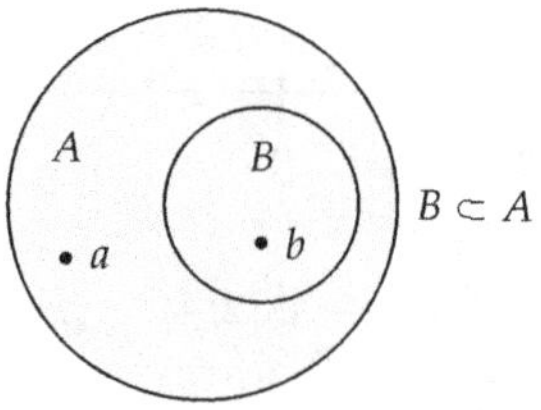

Figure 8.7: Venn diagram showing an example of the set relation $B \subset A$. The set B is strictly contained in the set A.

Figure 8.7 shows the picture that mathematicians have in mind when they say, "The set B is contained in the set A." The picture helps us visualize this abstract mathematical notion.

Mathematicians use two different symbols to describe set containment, in order to specify either a *strict* containment relation or a *subset-or-equal* relation. The two types of containment relations between sets are similar to the *less-than* ($<$) and *less-than-or-equal* ($\leqslant$) relations between numbers. A strict containment relation is denoted by the symbol $\subset$. We write $B \subset A$ if and only if every element of B is also an element of A, and there exists at least one element of A that is not an element of B. Using set notation, the previous sentence is expressed as

$$B \subset A \qquad \Leftrightarrow \qquad \forall b \in B, b \in A \text{ and } \exists a \in A \text{ such that } a \notin B.$$

For example, the expression $E \subset \mathbb{Z}$ shows that the even numbers are a strict subset of the integers. Every even number is an integer, but there exist integers that are not even (the odd numbers). Some mathematicians prefer the more descriptive symbol $\subsetneq$ to describe strict containment relations.

A subset-or-equal relation is denoted $B \subseteq A$. In writing $B \subseteq A$, a mathematician claims, "Every element of B is also an element of A," but makes no claim about the existence of elements that are contained in A but not in B. The statement $B \subset A$ implies $B \subseteq A$; however, $B \subseteq A$ does not imply $B \subset A$. This is analogous to how $b < a$ implies $b \leqslant a$, but $b \leqslant a$ doesn't imply $b < a$, since a and b could be equal.

Set operations

Venn diagrams also help us visualize the subsets obtained from set operations. Figure 8.8 illustrates the set union $A \cup B$, the set intersection $A \cap B$, and the set difference $A \backslash B$, for two sets A and B.

The union $A \cup B$ describes all elements that are in either set A or set B, or both. If $e \in A \cup B$, then $e \in A$ or $e \in B$.

Recall the set of even numbers $E \subset \mathbb{Z}$ and the set of odd numbers $O \subset \mathbb{Z}$ defined above. Since every integer is either an even number or an odd number, we know $\mathbb{Z} \subseteq E \cup O$. The union of two subsets is always contained within the parent set, so we also know $E \cup O \subseteq \mathbb{Z}$. Combining these facts, we can establish the equality $E \cup O = \mathbb{Z}$, which states the fact, "The combination of all even and odd numbers is the same as all integers."

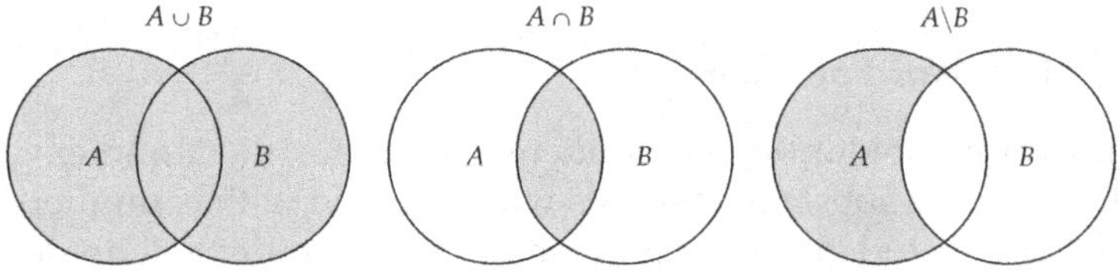

Figure 8.8: Venn diagrams showing different subsets obtained using the set operations: set union $A \cup B$, set intersection $A \cap B$, and set difference $A \backslash B$.

The set intersection $A \cap B$ and set difference $A \backslash B$ are also illustrated in Figure 8.8. The intersection of two sets contains the elements that are part of both sets. The set difference $A \backslash B$ contains all the elements that are in A but not in B.

Note the meaning of the conjunction "or" in English is ambiguous. The expression "in A or B" could be interpreted as either an "inclusive or," meaning "in A or B, or in both"—or as an "exclusive or," meaning "in A or B, but not both." Mathematicians always use "or" in the inclusive sense, so $A \cup B$ denotes elements that are in A or B, or in both sets. We can obtain an expression that corresponds to the "exclusive or" of two sets by taking the union of the sets and subtracting their intersection: $(A \cup B)\backslash(A \cap B)$.

Example 3: Set operations

Consider the three sets $A = \{a,b,c\}$, $B = \{b,c,d\}$, and $C = \{c,d,e\}$. Using set operations, we can define new sets, such as

$$A \cup B = \{a,b,c,d\}, \quad A \cap B = \{b,c\}, \quad \text{and} \quad A\backslash B = \{a\},$$

which correspond to elements in either A or B, the set of elements in A and B, and the set of elements in A but not in B, respectively.

We can also construct expressions involving three sets:

$$A \cup B \cup C = \{a,b,c,d,e\}, \qquad A \cap B \cap C = \{c\}.$$

And we can write more elaborate set expressions, like

$$(A \cup B)\backslash C = \{a,b\},$$

which is the set of elements that are in A or B but not in C.

Another example of a complicated set expression is

$$(A \cap B) \cup (B \cap C) = \{b,c,d\},$$

which describes the set of elements in both A and B or in both B and C. As you can see, set notation is a compact, precise language for writing complicated set expressions.

Example 4: Word problem

A startup is looking to hire student interns for the summer. Define C to be the subset of students who are good with computers, M the subset of students who know math, D the students with design skills, and L the students with good language skills.

Using set notation, we can specify different subsets of the students the startup might hire. Let's say the startup is a math textbook publisher; they want to hire students from the set $M \cap L$—the students who are good at math and who also have good language skills. A startup that builds websites needs both designers and coders, and therefore would choose students from the set $D \cup C$.

New vocabulary

The specialized notation used by mathematicians can be difficult to get used to. You must learn how to read symbols like $\exists$, $\subset$, $|$, and $\in$ and translate their meaning in the sentence. Indeed, learning advanced mathematics notation is akin to learning a new language.

To help you practice the new vocabulary, we'll look at a simple mathematical proof that makes use of the new symbols.

Simple proof example

Claim: Given $J(n) = 3n + 2 - n$, $J(n) \in E$ for all $n \in \mathbb{Z}$.

The claim is that the function $J(n)$ outputs an even number, whenever the input n is an integer. To prove this claim, we have to show that the expression $3n + 2 - n$ is even for all numbers $n \in \mathbb{Z}$.

Proof. We want to show $J(n) \in E$ for all $n \in \mathbb{Z}$. Let's first review the definition of the set of even numbers $E \stackrel{\text{def}}{=} \{m \in \mathbb{Z} \mid m = 2n, n \in \mathbb{Z}\}$. A number is even if it is equal to $2n$ for some integer n. Next let's simplify the expression for $J(n)$ as follows:

$$J(n) = 3n + 2 - n = 2n + 2 = 2(n + 1).$$

Observe that the number $(n + 1)$ is always an integer whenever n is an integer. Since the output of $J(n) = 2(n + 1)$ is equal to $2m$ for some integer m, we've proven that $J(n) \in E$, for all $n \in \mathbb{Z}$. $\square$

Sets as solutions to equations

Another context where sets come up is when describing solutions to equations and inequalities. In Section 1.1 we learned how to solve for the unknown x in equations. To solve the equation $f(x) = c$ is to find all the values of x that satisfy this equation. For simple equations like $x - 3 = 6$, the solution is a single number $x = 9$, but more complex equations can have multiple solutions. For example, the solution to the equation $x^2 = 4$ is the set $\{-2, 2\}$, since both $x = -2$ and $x = 2$ satisfy the equation.

Please update your definition of the math verb "to solve" (an equation) to include the new notion of a *solution set*—the set of values that satisfy the equation. A solution set is the mathematically precise way to describe an equation's solutions:

- The solution set to the equation $x - 3 = 6$ is the set $\{9\}$.
- The solution set for the equation $x^2 = 4$ is the set $\{-2, 2\}$.
- The solution set of $\sin(x) = 0$ is the set $\{x \mid x = \pi n, \forall n \in \mathbb{Z}\}$.

- The solution set for the equation $\sin(x) = 2$ is $\varnothing$ (the empty set), since there is no number x that satisfies the equation.

The SymPy function solve returns the solutions of equations as a list. To solve the equation $f(x) = c$ using SymPy, we first rewrite it as expression that equals zero $f(x) - c = 0$, then call the function solve:

```
>>> solve(x-3 -6, x)           # usage: solve(expr, var)
[9]

>>> solve(x**2 -4, x)
[-2, 2]

>>> solve(sin(x), x)
[0, pi]                        # found only solutions in [0,2*pi)

>>> solve(sin(x) -2, x)
[]                             # empty list = empty set
```

In the next section we'll learn how the notion of a solution set is used for describing the solutions to systems of equations.

Solution sets to systems of equations

Let's revisit what we learned in Section 8.1 about the solutions to systems of linear equations, and define their solution sets more precisely. The solution set for the system of equations

$$\begin{aligned} a_1x + b_1y &= c_1, \\ a_2x + b_2y &= c_2, \end{aligned}$$

corresponds to the intersection of two sets:

$$\underbrace{\{(x,y) \in \mathbb{R}^2 \mid a_1x + b_1y = c_1\}}_{\ell_1} \cap \underbrace{\{(x,y) \in \mathbb{R}^2 \mid a_2x + b_2y = c_2\}}_{\ell_2}.$$

Recall that the lines ℓ_1 and ℓ_2 are the geometric interpretation of these sets. Each line corresponds to a set of coordinate pairs (x, y) that satisfy the equation of the line. The solution to the system of equations is the set of points at the intersection of the two lines $\ell_1 \cap \ell_2$. Note the word *intersection* is used in two different mathematical contexts: the solution is the *set intersection* of two sets, and also the *geometric intersection* of two lines.

Let's take advantage of this correspondence between set intersections and geometric line intersections to understand the solutions to systems of equations in a little more detail. In the next three sections, we'll look at three possible cases that can occur when trying to solve a system of two linear equations in two unknowns. So far we've only discussed Case A, which occurs when the two lines intersect at a point, as in the example shown in Figure 8.9. To fully understand the possible solutions to a system of equations, we need to think about all other cases; like Case B when $\ell_1 \cap \ell_2 = \emptyset$ as in Figure 8.10, and Case C when $\ell_1 \cap \ell_2 = \ell_1 = \ell_2$ as in Figure 8.11.

Case A: One solution. When the lines ℓ_1 and ℓ_2 are non-parallel, they will intersect at a point as shown in Figure 8.9. In this case, the solution set to the system of equations contains a single point:

$$\{(x,y) \in \mathbb{R}^2 \mid x + 2y = 2\} \cap \{(x,y) \in \mathbb{R}^2 \mid x = 1\} = \{(1, \tfrac{1}{2})\}.$$

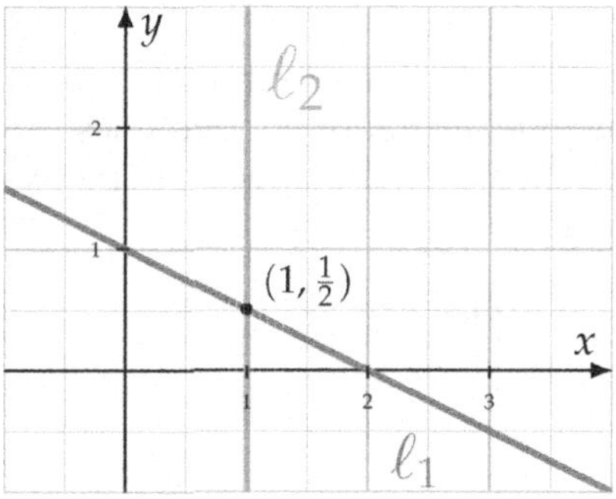

Figure 8.9: Case A: The intersection of the lines with equations $x + 2y = 2$ and $x = 1$ is the point $(1, \frac{1}{2}) \in \mathbb{R}^2$.

Case B: No solution. If the lines ℓ_1 and ℓ_2 are parallel then they will never intersect. The intersection of these lines is the empty set:

$$\{(x,y) \in \mathbb{R}^2 \mid x + 2y = 2\} \cap \{(x,y) \in \mathbb{R}^2 \mid x + 2y = 4\} = \emptyset.$$

Think about it—there is no point (x, y) that lies on both ℓ_1 and ℓ_2. Using algebra terminology, we say this system of equations has no solution, since there are no numbers x and y that satisfy both equations.

Case C: Infinitely many solutions. If the lines ℓ_1 and ℓ_2 are parallel and overlapping then they intersect everywhere. This case occurs when one of the equations in a system of equations is a multiple of the other equation, as in the case of equations $x + 2y = 2$ and $3x + 6y = 6$. The lines ℓ_1 and ℓ_2 that correspond to these equations are shown in Figure 8.11. Any point (x, y) that satisfies $x + 2y = 2$

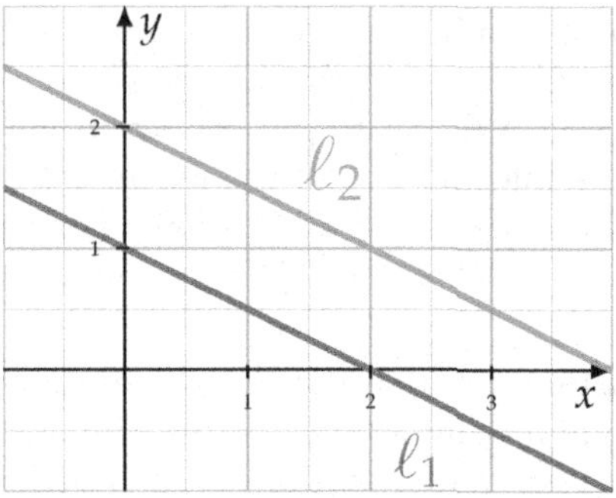

Figure 8.10: Case B: The lines with equations $x + 2y = 2$ and $x + 2y = 4$ are parallel and do not intersect. Using set notation, we can describe the solution set as $\varnothing$ (the empty set).

also satisfies $3x + 6y = 6$. Since both equations describe the same geometric line, the intersection of the two lines is equal to the lines: $\ell_1 \cap \ell_2 = \ell_1 = \ell_2$. In this case, the solution to the system of equations is described by the set $\{(x, y) \in \mathbb{R}^2 \mid x + 2y = 2\}$.

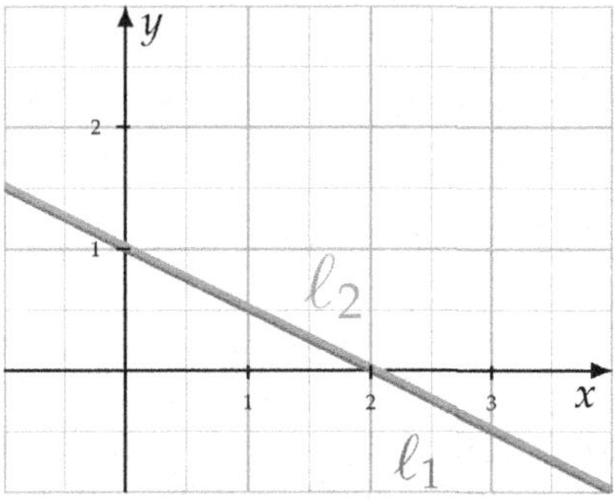

Figure 8.11: Case C: the line ℓ_1 described by equation $x + 2y = 2$ and the line ℓ_2 described by equation $3x + 6y = 6$ correspond to the same line in the Cartesian plane. The intersection of these lines is the set $\{(x, y) \in \mathbb{R}^2 \mid x + 2y = 2\} = \ell_1 = \ell_2$.

We need to consider all three cases when thinking about the solutions to systems of linear equations: the solution set can be a point (Case A), the empty set (Case B), or a line (Case C). Observe that the same mathematical notion (a set) is able to describe the solutions in all three cases even though the solutions correspond to very different geometric objects. In Case A the solution is a set that contains a single point $\{(x, y)\}$. In Case B the solution is the empty set $\varnothing$. And in Case C the solution set is described by the infinite set $\{(x, y) \in \mathbb{R}^2 \mid ax + by = c\}$, which corresponds to a line ℓ in the

Cartesian plane. I hope you'll agree with me that set notation is useful for describing mathematical concepts precisely and for handling solutions to linear equations.

Sets are also useful for describing the solutions to inequalities, which is what we'll learn about next.

Inequalities

In this section, we'll learn how to solve inequalities. The solution set to an inequality is an *interval*—a subset of the number line. Consider the inequality $x^2 \leqslant 4$, which is equivalent to asking the question, "For which values of x is x^2 less than or equal to 4?" The answer to this question is the interval $[-2, 2] = \{x \in \mathbb{R} \mid -2 \leqslant x \leqslant 2\}$.

Working with inequalities is essentially the same as working with their endpoints. To solve the inequality $x^2 \leqslant 4$, we first solve $x^2 = 4$ to find the endpoints and then use trial and error to figure out which part of the space to the left and right of the endpoints satisfies the inequality.

It's important to distinguish the different types of inequality conditions. The four different types of inequalities are

- $f(x) < g(x)$: a strict inequality. The function $f(x)$ is always *strictly less than* the function $g(x)$.
- $f(x) \leqslant g(x)$: the function $f(x)$ is *less than or equal to* $g(x)$.
- $f(x) > g(x)$: $f(x)$ is *strictly greater than* $g(x)$.
- $f(x) \geqslant g(x)$: $f(x)$ is *greater than or equal to* $g(x)$.

Depending on the type of inequality, the answer will be either a *open* or *closed* interval.

To solve inequalities we use the techniques we learned for solving equations: we perform simplifying steps **on both sides of the inequality** until we obtain the answer. The only new aspect when dealing with inequalities is the following. When multiplying an inequality by a negative number on both sides, we must flip the direction of the inequality:

$$f(x) \leqslant g(x) \qquad \Rightarrow \qquad -f(x) \geqslant -g(x).$$

Example 5 To solve the inequality $7 - x \leqslant 5$ we must *dig* toward the x and *undo* all the operations that stand in our way:

$$\begin{aligned} 7 - x &\leqslant 5, \\ (-x) + 7 &\leqslant 5, \\ (-x) + 7 - 7 &\leqslant 5 - 7, \\ -x &\leqslant -2, \\ x &\geqslant 2. \end{aligned}$$

To obtain the second line we simply rewrote the order of operations on the left side of the inequality. In the third line we subtracted 7 from both sides of the inequality to undo the $+7$ operation. In the last step we multiplied both sides of the inequality by -1, which had the effect of changing the inequality from $\leqslant$ to $\geqslant$. The solution set to the inequality $7 - x \leqslant 5$ is the interval $[2, \infty)$.

Example 6 To solve the inequality $x^2 \leqslant 4$, we must undo the quadratic function by taking the square root of both sides of the inequality. Note the equation $x^2 = 4$ has two solutions: $x = -2$ and $x = 2$. Similarly, we'll need to consider two separate cases for the inequality conditions. Simplifying the inequality $x^2 \leqslant 4$ by taking the square root on both sides results in two inequality conditions

$$x \geqslant -2 \qquad \text{and} \qquad x \leqslant 2,$$

which we can express more concisely as $-2 \leqslant x \leqslant 2$. If x is a negative number, it must be greater than -2; and if x is a positive number, it must be less than 2 in order for $x^2 \leqslant 4$. The solution set for the inequality $x^2 \leqslant 4$ is the interval $[-2, 2] = \{x \in \mathbb{R} \mid -2 \leqslant x \leqslant 2\}$. Note the solution is a closed interval (square brackets), which means the endpoints are included.

The best way to convince yourself that the above algebraic reasoning is correct is to think about the graph of the function $f(x) = x^2$. The inequality $x^2 \leqslant 4$ corresponds to the condition $f(x) \leqslant 4$. For what values of x is the graph of the function $f(x)$ below the line with equation $y = 4$?

As you can see, solving inequalities is no more complicated than solving equations. You can think about an inequality in terms of its endpoints, which correspond to the equality conditions. Whenever things get complicated (as in Example 6), you can sketch the function graphs for the different terms in the inequality and visually determine the appropriate directions for the inequality signs.

Sets related to functions

A function that takes real variables as inputs and produces real numbers as outputs is denoted $f : \mathbb{R} \to \mathbb{R}$. The *domain* of a function is the set of all possible inputs to the function that produce an output:

$$\text{Dom}(f) \stackrel{\text{def}}{=} \{x \in \mathbb{R} \mid f(x) \in \mathbb{R}\}.$$

Inputs for which the function is undefined are not part of the domain. For instance the function $f(x) = \sqrt{x}$ is not defined for negative inputs, so we have $\text{Dom}(f) = \mathbb{R}_+$.

The *image* of a function is the set of all possible outputs of the function:

$$\text{Im}(f) \stackrel{\text{def}}{=} \{y \in \mathbb{R} \mid \exists x \in \mathbb{R},\ y = f(x)\}.$$

For example, the function $f(x) = x^2$ has the image set $\text{Im}(f) = \mathbb{R}_+$ since the outputs it produces are always nonnegative.

Discussion

Knowledge of the precise mathematical jargon introduced in this section is not crucial to understanding basic mathematics. That said, I wanted to expose you to some technical math notation here because this is the language in which mathematicians think and communicate. Most advanced math textbooks will assume you understand technical math notation, so it's good to be prepared.

Exercises

E8.7 Given the three sets $A = \{1,2,3,4,5,6,7\}$, $B = \{1,3,5\}$, and $C = \{2,4,6\}$, compute the following set expressions.

a) $A \backslash B$ **b)** $B \cup C$ **c)** $A \cap B$ **d)** $B \cap C$

e) $A \cup B \cup C$ **f)** $A \backslash (B \cup C)$ **g)** $(A \backslash B) \cup C$ **h)** $A \cap B \cap C$

E8.8 Find the values of x that satisfy the following inequalities.

a) $2x < 3$ **b)** $-4x \geqslant 20$ **c)** $|2x - 3| < 5$

d) $3x + 3 < 5x - 5$ **e)** $\frac{1}{2}x - 2 \geqslant \frac{1}{3}$ **f)** $(x + 1)^2 \geqslant 9$

Express your answer as an interval with appropriate endpoints.

Chapter 9

Practice problems

We've now reached the problems chapter in this book. The purpose of these problems is to give you a way to comprehensively practice your math fundamentals. Knowing how to solve math problems is a very useful skill to develop. At times, honing your math chops might seem like tough mental work, but at the end of each problem, you'll gain a stronger foothold on all the topics you've been learning about. You'll also experience a small *achievement buzz* after each problem you vanquish.

Sit down and take a crack at these practice problems today, or another time when you're properly caffeinated. If you make time for some math practice, you'll develop long-lasting comprehension and true math fluency.

Without solving any problems, you're likely to forget most of what you've learned in the next few months. You might still remember the big ideas, but the details will be fuzzy and faded. By solving some of the practice problems, you'll remember a lot more stuff. Don't break the pace now: with math, it's very much *use it or lose it*!

Make sure you put your phone away while you're working on the problems. You don't need fancy technology to do math; grab a pen and some paper from the printer and you'll be fine. The great mathematicians like Descartes, Hilbert, Leibniz, and Noether did most of their work with pen and paper and they did well. Spend some time with math the way they did.

P9.1 Solve for x in the equation $x^2 - 9 = 7$.

P9.2 Solve for x in the equation $\cos^{-1}\left(\frac{x}{A}\right) - \phi = \omega t$.

P9.3 Solve for x in the equation $\frac{1}{x} = \frac{1}{a} + \frac{1}{b}$.

P9.4 Use a calculator to find the values of the following expressions:

a) $\sqrt[4]{3^3}$ **b)** 2^{10} **c)** $7^{\frac{1}{4}} - 10$ **d)** $\frac{1}{2}\ln(e^{22})$

P9.5 Compute the following expressions involving fractions:

a) $\frac{1}{2} + \frac{1}{4}$ **b)** $\frac{4}{7} - \frac{23}{5}$ **c)** $1\frac{3}{4} + 1\frac{31}{32}$

P9.6 Use the basic rules of algebra to simplify the following expressions:

a) $ab\,\frac{1}{a}\,b^2cb^{-3}$ **b)** $\frac{abc}{bca}$ **c)** $\frac{27a^2}{\sqrt{9abba}}$

d) $\frac{a(b+c)-ca}{b}$ **e)** $\frac{a}{c\sqrt[3]{b}}\frac{b^{\frac{4}{3}}}{a^2}$ **f)** $(x+a)(x+b) - x(a+b)$

P9.7 Expand the brackets in the following expressions:

a) $(x+a)(x-b)$ **b)** $(2x+3)(x-5)$ **c)** $(5x-2)(2x+7)$

P9.8 Factor the following expressions as a product of linear terms:

a) $x^2 - 2x - 8$ **b)** $3x^3 - 27x$ **c)** $6x^2 + 11x - 21$

P9.9 Complete the square in the following quadratic expressions to obtain expressions of the form $A(x-h)^2 + k$.

a) $x^2 - 4x + 7$ **b)** $2x^2 + 12x + 22$ **c)** $6x^2 + 11x - 21$

P9.10 A golf club and a golf ball cost \$1.10 together. The golf club costs one dollar more than the ball. How much does the ball cost?

P9.11 An ancient artist drew scenes of hunting on the walls of a cave, including 43 figures of animals and people. There were 17 more figures of animals than people. How many figures of people did the artist draw and how many figures of animals?

P9.12 A father is 35 years old and his son is 5 years old. In how many years will the father's age be four times the son's age?

P9.13 A boy and a girl collected 120 nuts. The girl collected twice as many nuts as the boy. How many nuts did each collect?

P9.14 Alice is 5 years older than Bob. The sum of their ages is 25 years. How old is Alice?

P9.15 A publisher needs to bind 4500 books. One print shop can bind these books in 30 days, another shop can do it in 45 days. How many days are necessary to bind all the books if both shops work in parallel?
Hint: Find the books-per-day rate of each shop.

P9.16 A plane leaves Vancouver travelling at 600 km/h toward Montreal. One hour later, a second plane leaves Vancouver heading for Montreal at 900 km/h. How long will it take for the second plane to overtake the first?
Hint: Distance travelled is equal to velocity multiplied by time: $d = vt$.

P9.17 There are 26 sheep and 10 goats on a ship. How old is the captain?

P9.18 The golden ratio, denoted φ, is the positive solution to the quadratic equation $x^2 - x - 1 = 0$. Find the golden ratio.

P9.19 Solve for x in the equation $\frac{1}{x} + \frac{2}{1-x} = \frac{4}{x^2}$.
Hint: Multiply both sides of the equation by $x^2(1-x)$.

P9.20 Use substitution to solve for x in the following equations:

a) $x^6 - 4x^3 + 4 = 0$ **b)** $\frac{1}{2-\sin x} = \sin x$

P9.21 Find the range of values of the parameter m for which the equation $2x^2 - mx + m = 0$ has no real solutions.
Hint: Use the quadratic formula.

P9.22 Use the properties of exponents and logarithms to simplify

a) $e^x e^{-x} e^z$ **b)** $\left(\frac{xy^{-2}z^{-3}}{x^2y^3z^{-4}}\right)^{-3}$ **c)** $(8x^6)^{-\frac{2}{3}}$

d) $\log_4(\sqrt{2})$ **e)** $\log_{10}(0.001)$ **f)** $\ln(x^2-1) - \ln(x-1)$

P9.23 When representing numbers on a computer, the number of digits of precision n in base b and the approximation error ϵ are related by the equation $n = -\log_b(\epsilon)$. A `float64` has 53 bits of precision (digits base 2). What is the approximation error ϵ for a `float64`? How many digits of precision does a `float64` have in decimal (base 10)?

P9.24 Find the values of x that satisfy the following inequalities:

a) $2x - 5 > 3$ **b)** $5 \leqslant 3x - 4 \leqslant 14$ **c)** $2x^2 + x \geqslant 1$

P9.25 Two algorithms, P and Q, can be used to solve a certain problem. The running time of Algorithm P as a function of the size of the problem n is described by the function $P(n) = 0.002n^2$. The running time of Algorithm Q is described by $Q(n) = 0.5n$. For small problems, Algorithm P runs faster. Starting from what n will Algorithm Q be faster?

P9.26 Consider a right-angle triangle in which the shorter sides are 8 cm and 6 cm. What is the length of the triangle's longest side?

P9.27 A television screen measures 26 inches on the diagonal. The screen height is 13 inches. How wide is the screen?

P9.28 A ladder of length 3.33 m leans against a wall and its foot is 1.44 m from the wall. What is the height h where the ladder touches the wall?

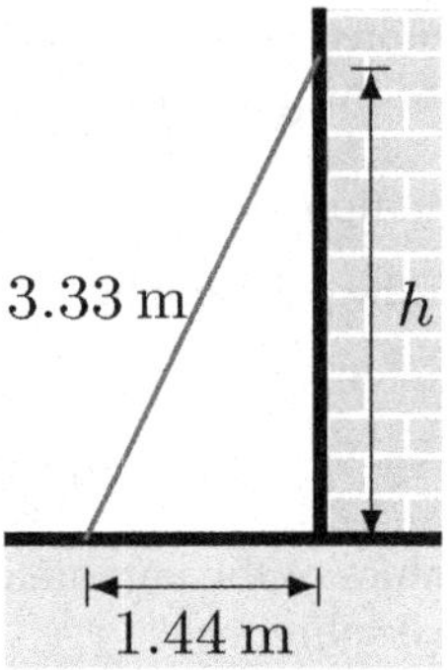

P9.29 **Kepler's triangle** Consider a right-angle triangle in which the hypotenuse has length $\varphi = \frac{\sqrt{5}+1}{2}$ (the golden ratio) and the adjacent side has length $\sqrt{\varphi}$. What is the length of the opposite side?

P9.30 Find the lengths x, y, and z in the figure below.

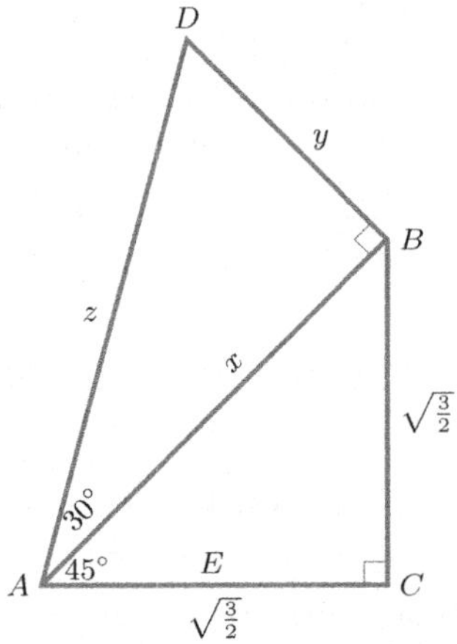

P9.31 Given the angle and distance measurements labelled in Figure 9.1, calculate the distance d and the height of the mountain peak h.

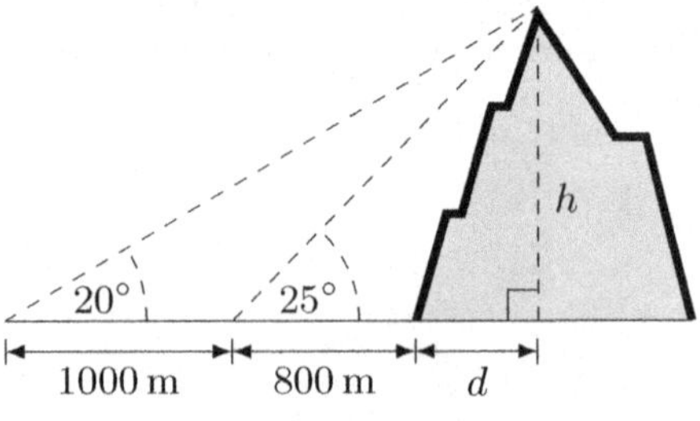

Figure 9.1: Measuring the height of a mountain using angles.

Hint: Use the definition of $\tan\theta$ to obtain two equations in two unknowns.

P9.32 You're observing a house from a blimp flying at an altitude of 2000 metres. From your point of view, the house appears at an angle $24°$ below the horizontal. What is the horizontal distance x between the blimp and the house?

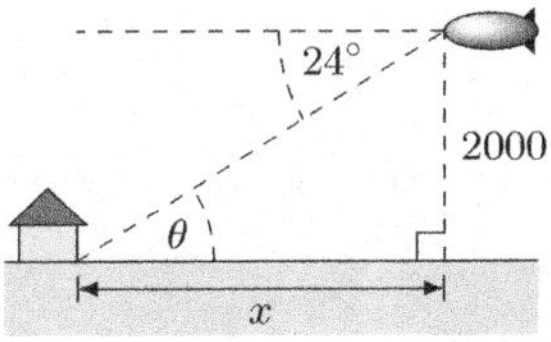

P9.33 Find x. Express your answer in terms of a, b, c and θ.

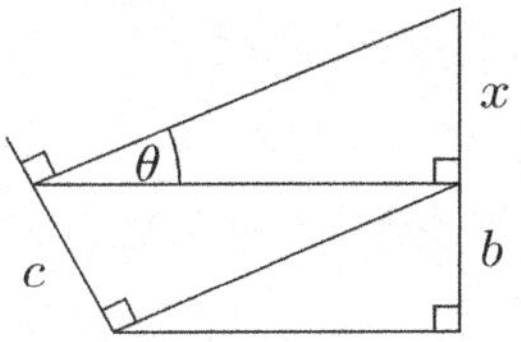

Hint: Use Pythagoras' theorem twice; then use the function tan.

P9.34 An equilateral triangle is inscribed in a circle of radius 1. Find the side length a and the area of the inscribed triangle $A_{\triangle}$.

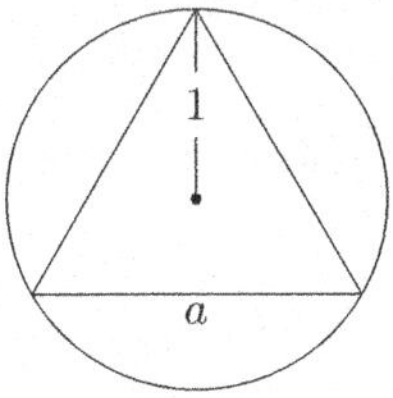

Hint: Split the triangle into three sub-triangles.

P9.35 Use the power-reduction trigonometric identities (page 106) to express $\sin^2\theta\cos^2\theta$ in terms of $\cos 4\theta$.

P9.36 A circle of radius 1 is inscribed inside a *regular octagon* (a polygon with eight sides of length b). Calculate the octagon's perimeter and its area.

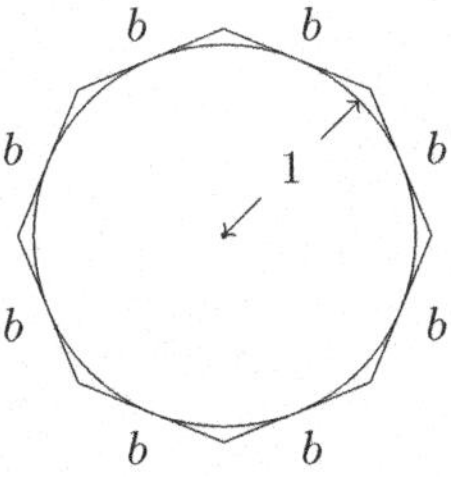

Hint: Split the octagon into eight isosceles triangles.

P9.37 Find the length of side c in the triangle:

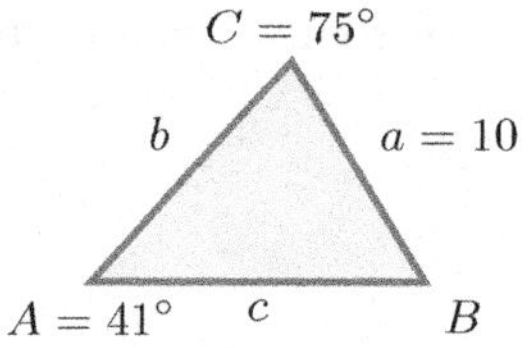

Hint: Use the sine rule.

P9.38 Consider the obtuse triangle shown in Figure 9.2.

a) Express h in terms of a and θ.

b) What is the area of this triangle?

c) Express c in terms of the variables a, b, and θ.

Hint: You can use the cosine rule for part **c)**.

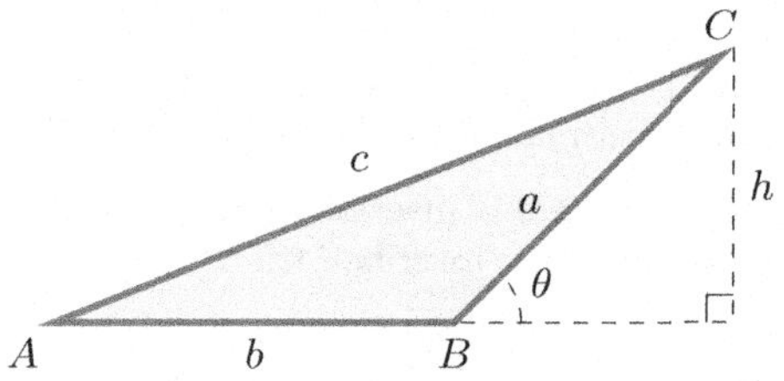

Figure 9.2: A triangle with base b and height h.

P9.39 Find the measure of the angle B and deduce the measure of the angle C. Find the length of side c.

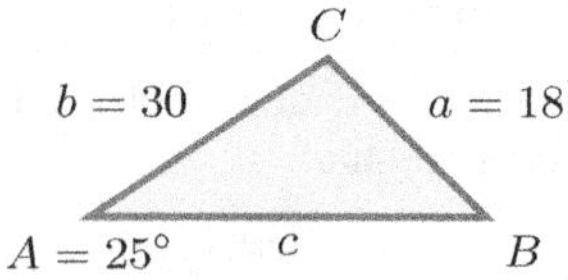

Hint: The sum of the internal angle measures of a triangle is $180°$.

P9.40 An observer on the ground measures an angle of inclination of $30°$ to an approaching airplane, and 10 seconds later measures an angle of inclination of $55°$. If the airplane is flying at a constant speed at an altitude of 2000 m in a straight line directly over the observer, find the speed of the airplane in kilometres per hour.

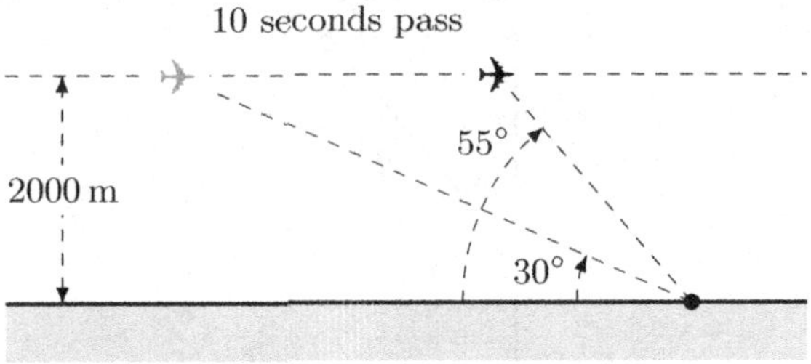

P9.41 Satoshi likes warm saké. He places 1 litre of water in a sauce pan with diameter 17 cm. How much will the height of the water level rise when Satoshi immerses a saké bottle with diameter 7.5 cm?
Hint: You'll need the volume conversion ratio 1 litre = 1000 cm^3.

P9.42 Find the length x of the diagonal of the quadrilateral below.

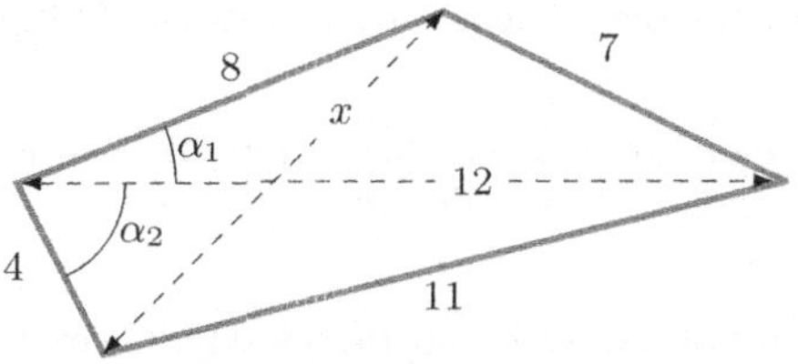

Hint: Use the law of cosines once to find α_1 and α_2, and again to find x.

P9.43 Find the area of the shaded region.

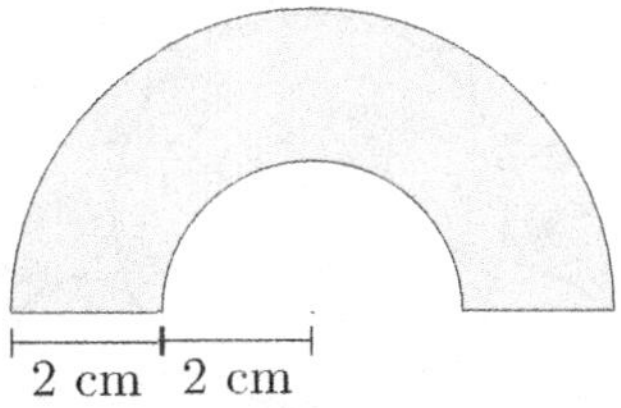

Hint: Find the area of the outer circle, subtract the area of missing centre disk, then divide by two.

P9.44 In preparation for the shooting of a music video, you're asked to suspend a wrecking ball hanging from a circular pulley. The pulley has a radius of 50 cm. The other lengths are indicated in the figure. What is the total length of the rope required?

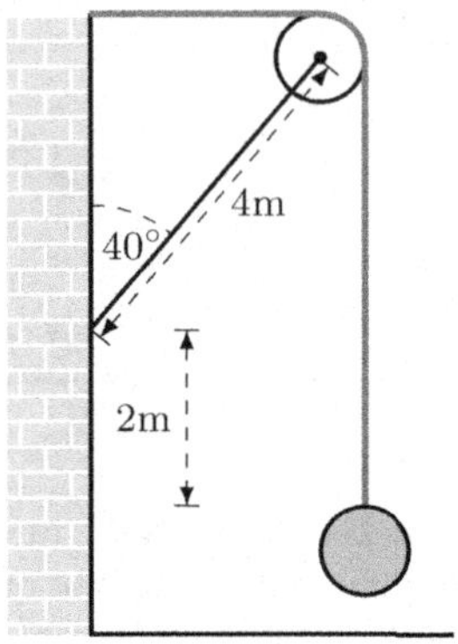

Hint: The total length of rope consists of two straight parts and the curved section that wraps around the pulley.

P9.45 The length of a rectangle is $c + 2$ and its height is 5. What is the area of the rectangle?

P9.46 A box of facial tissues has dimensions 10.5 cm by 7 cm by 22.3 cm. What is the volume of the box in litres?
Hint: $1\,\text{L} = 1000\,\text{cm}^3$.

P9.47 A large circle of radius R is surrounded by 12 smaller circles of radius r. Find the ratio $\frac{R}{r}$ rounded to four decimals.

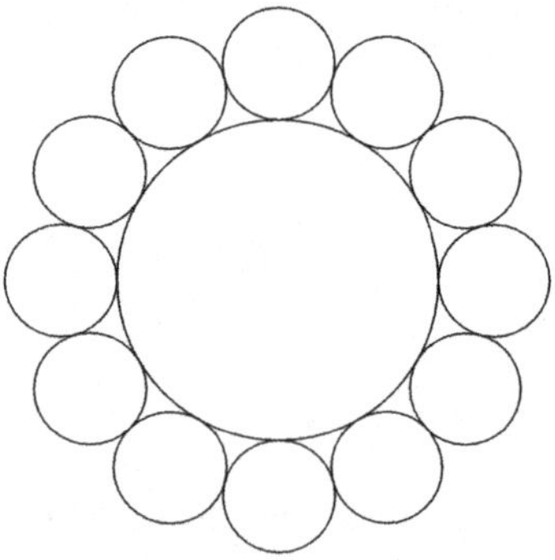

Hint: Draw an isosceles triangle with one vertex at the centre of the R-circle and the other vertices at the centres of two adjacent r-circles.

P9.48 The area of a rectangular figure is $35\,\text{cm}^2$. If one side is 5 cm, how long is the other side?

P9.49 A swimming pool has length $\ell = 20\,\text{m}$, width $w = 10\,\text{m}$, and depth $d = 1.5\,\text{m}$. Calculate the volume of water in the swimming pool in litres?
Hint: $1\,\text{m}^3 = 1000\,\text{L}$.

P9.50 How many litres of water remain in a tank that is 15 m long, 6 m wide, and 5 m high, if 30% of its capacity is spent?

P9.51 A building has two water tanks, each with capacity 4000 L. One of them is $\frac{1}{4}$ full and the other contains three times more water. How many litres of water does the building have in total?

P9.52 The rectangular lid of a box has length 40 cm and width 30 cm. A rectangular hole with area 500 cm^2 must be cut in the lid so that the hole's sides are equal distances from the sides of the lid. What will the distance be between the sides of the hole and the sides of the lid?
Hint: You'll need to define three variables to solve this problem.

P9.53 A man sells firewood. To make standard portions, he uses a standard length of rope ℓ to surround a pack of logs. One day, a customer asks him for a double portion of firewood. What length of rope should he use to measure this order? Assume the packs of logs are circular in shape.

P9.54 How much pure water should be added to 10 litres of a solution that is 60% acid to make a solution that is 20% acid?

P9.55 A tablet screen has a resolution of 768 pixels by 1024 pixels, and the physical dimensions of the screen are 6 inches by 8 inches. One might conclude that the best size of a PDF document for such a screen would be 6 inches by 8 inches. At first I thought so too, but I forgot to account for the status bar, which is 20 pixels tall. The actual usable screen area is only 768 pixels by 1004 pixels. Assuming the width of the PDF is chosen to be 6 inches, what height should the PDF be so that it fits perfectly in the content area of the tablet screen?

P9.56 Find the sum of the natural numbers 1 through 100.
Hint: Imagine pairing the biggest number with the smallest number in the sum, the second biggest number with the second smallest number, etc.

P9.57 Express the following vectors in length-and-direction notation:

a) $\vec{u}_1 = (0, 5)$ **b)** $\vec{u}_2 = (1, 2)$ **c)** $\vec{u}_3 = (-1, -2)$

P9.58 Express the following vectors as components:

a) $\vec{v}_1 = 20\angle 30°$ **b)** $\vec{v}_2 = 10\angle -90°$ **c)** $\vec{v}_3 = 5\angle 150°$

P9.59 Express the following vectors in terms of unit vectors $\hat{\imath}$, $\hat{\jmath}$, and $\hat{k}$:

a) $\vec{w}_1 = 10\angle 25°$ **b)** $\vec{w}_2 = 7\angle -90°$ **c)** $\vec{w}_3 = (3, -2, 3)$

P9.60 Given the vectors $\vec{v}_1 = (1, 1)$, $\vec{v}_2 = (2, 3)$, and $\vec{v}_3 = 5\angle 30°$, calculate the following expressions:

a) $\vec{v}_1 + \vec{v}_2$ **b)** $\vec{v}_2 - 2\vec{v}_1$ **c)** $\vec{v}_1 + \vec{v}_2 + \vec{v}_3$

P9.61 Starting from the point $P = (2, 6)$, the three displacement vectors shown in Figure 9.3 are applied to obtain the point Q. What are the coordinates of the point Q?

P9.62 Given the vectors $\vec{u} = (1, 1, 1)$, $\vec{v} = (2, 3, 1)$, and $\vec{w} = (-1, -1, 2)$, compute the following products:

a) $\vec{u} \cdot \vec{v}$ **b)** $\vec{u} \cdot \vec{w}$ **c)** $\vec{v} \cdot \vec{w}$

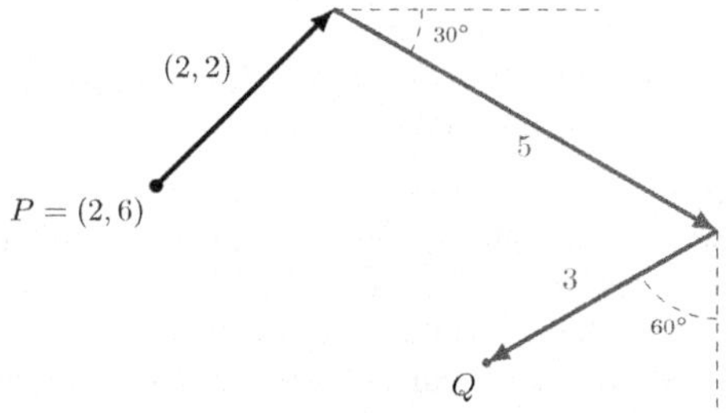

Figure 9.3: A point P is displaced by three vectors to obtain point Q.

d) $\vec{u} \times \vec{v}$ **e)** $\vec{u} \times \vec{w}$ **f)** $\vec{v} \times \vec{w}$

P9.63 Given the vectors $\vec{p} = (1,1,0,3,3)$ and $\vec{q} = (1,2,3,4,5)$, calculate the following expressions:

a) $\vec{p} + \vec{q}$ **b)** $\vec{p} - \vec{q}$ **c)** $\vec{p} \cdot \vec{q}$

P9.64 Find a unit vector that is perpendicular to both $\vec{u} = (1,0,1)$ and $\vec{v} = (1,2,0)$.
Hint: Use the cross product.

P9.65 Find a vector that is orthogonal to both $\vec{u}_1 = (1,0,1)$ and $\vec{u}_2 = (1,3,0)$, and whose dot product with the vector $\vec{v} = (1,1,0)$ is equal to 8.

P9.66 Compute the following expressions:

a) $\sqrt{-4}$ **b)** $\dfrac{2+3i}{2+2i}$ **c)** $e^{3i}(2+i)e^{-3i}$

P9.67 Solve for $x \in \mathbb{C}$ in the following equations:

a) $x^2 = -4$ **b)** $\sqrt{x} = 4i$

c) $x^2 + 2x + 2 = 0$ **d)** $x^4 + 4x^2 + 3 = 0$

Hint: To solve **d)**, use the substitution $u = x^2$.

P9.68 Given the numbers $z_1 = 2 + i$, $z_2 = 2 - i$, and $z_3 = -1 - i$, compute

a) $|z_1|$ **b)** $\dfrac{z_1}{z_3}$ **c)** $z_1 z_2 z_3$

P9.69 A real business is a business that is profitable. An imaginary business is an idea that is just turning around in your head. We can model the real-imaginary nature of a business project by representing the *project state* as a complex number $p \in \mathbb{C}$. For example, a business idea is described by the state $p_o = 100i$. In other words, it is 100% imaginary.

To bring an idea from the imaginary into the real, you must work on it. We'll model the work done on the project as a multiplication by the complex number $e^{-i\alpha h}$, where h is the number of hours of work and α is a constant that depends on the project. After h hours of work, the initial state of the project p_o has become $p_f = e^{-i\alpha h} p_o$. Working on the project for one hour "rotates" its state by $-\alpha$ rad, making it more real and less imaginary.

If you start from an idea $p_o = 100i$ and the cumulative number of hours invested after t weeks of working on the project is $h(t) = 0.2t^2$, how long will it take for the project to become 100% real? Assume $\alpha = 2.904 \times 10^{-3}$.
Hint: A project is 100% real if $\text{Re}\{p\} = p$.

P9.70 A farmer with a passion for robotics has built a prototype of a robotic tractor. The tractor is programmed to move with a speed of 0.524 km/h and follow the direction of the hour-hand on a conventional watch. Assume the tractor starts at 12:00 p.m. (noon) and is left to roam about in a field until 6:00 p.m. What is the shape of the trajectory that the tractor will follow? What is the total distance travelled by the tractor after six hours?

P9.71 Solve for x and y simultaneously in the following system of equations: $-x - 2y = -2$ and $3x + 3y = 0$.

P9.72 Solve the following system of equations for the three unknowns:

$$\begin{aligned} 1x + 2y + 3z &= 14, \\ 2x + 5y + 6z &= 30, \\ -1x + 2y + 3z &= 12. \end{aligned}$$

P9.73 A hotel offers a 15% discount on rooms. Determine the original price of a room if the discounted room price is \$95.20.

P9.74 A set of kitchen tools normally retails for \$450, but today it is priced at the special offer of \$360. Calculate the percentage of the discount.

P9.75 You take out a \$5000 loan at a nominal annual percentage rate (nAPR) of 12% with monthly compounding. How much money will you owe after 10 years?

P9.76 Plot the graphs of $f(x) = 100e^{-x/2}$ and $g(x) = 100(1 - e^{-x/2})$ by evaluating the functions at different values of x from 0 to 11.

P9.77 Starting from an initial quantity Q_o of Exponentium at $t = 0$ s, the quantity Q of Exponentium as a function of time varies according to the expression $Q(t) = Q_o e^{-\lambda t}$, where $\lambda = 5.0$ and t is measured in seconds. Find the *half-life* of Exponentium, that is, the time it takes for the quantity of Exponentium to reduce to half the initial quantity Q_o.

P9.78 A hot body cools so that every 24 min its temperature decreases by a factor of two. Deduce the time-constant and determine the time it will take the body to reach 1% of its original temperature.
Hint: The temperature function is $T(t) = T_o e^{-t/\tau}$ and τ is the *time constant*.

P9.79 A capacitor of capacitance $C = 4.0 \times 10^{-6}$ farads, charged to an initial potential of $V_o = 20$ volts, is discharging through a resistance of $R = 10\,000\,\Omega$ (read Ohms). Find the potential V after 0.01 s and after 0.1 s, knowing the decreasing potential follows the rule $V(t) = V_o e^{-\frac{t}{RC}}$.

P9.80 Let B be the set of people who are bankers and C be the set of crooks. Rewrite the math statement $\exists b \in B \mid b \notin C$ in plain English.

P9.81 Let M denote the set of people who run Monsanto, and H denote the people who ought to burn in hell for all eternity. Write the math statement $\forall p \in M, p \in H$ in plain English.

P9.82 When starting a business, one sometimes needs to find investors. Define M to be the set of investors with money, and C to be the set of investors with connections. Describe the following sets in words: **a)** $M \backslash C$, **b)** $C \backslash M$, and the most desirable set **c)** $M \cap C$.

P9.83 Write the formulas for the functions $A_1(x)$ and $A_2(x)$ that describe the areas of the following geometric shapes.

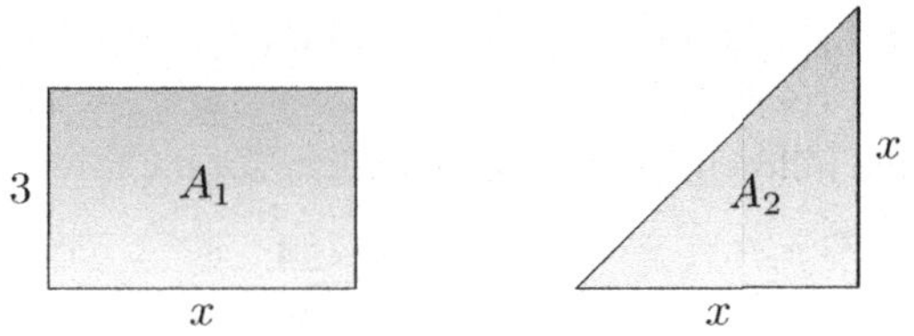

End matter

Conclusion

We covered many fundamental concepts during this hike in the math mountains. We started with basic math ideas like variables, expressions, and equations, and then moved into algebra—hopefully you'll agree with me that algebra is not as scary as some people think, right? We covered many other scary topics like exponents and logarithms, and you made it through unscathed. We even tackled more advanced topics like functions, vectors, complex numbers, proofs, and set notation. Congratulations on powering through this intensive math bootcamp!

The only way to truly understand math is to apply math concepts to solving problems. You don't get points for memorizing rules, formulas, and equations; only when you learn how to *use* the equations will you truly "own" the math. Math is not a spectator sport! If you're reading this book without solving the practice problems, you're getting a bad deal. If you haven't attempted to solve the problems in Chapter 9, you need to turn back and give them a shot.

My intention with this book was to give you a solid, basic understanding of many mathematical concepts, explained quickly and simply so you wouldn't get bored. However, it's possible that the book's fast pace led to some gaps in the coverage of the material. Let me know if you find any parts of the book that lack clarity or are missing explanations. Send me an email if you find a confusing explanation, a missing definition, or an equation you think is wrong. You can reach me at `ivan@minireference.com`. The book's relative errorlessness is thanks to the many typo fixes, bug reports, and helpful suggestions I receive from readers like you.

To learn more about other books in the **No Bullshit Guide** series and find out how Minireference Co. is revolutionizing the textbook industry, visit our blog at `minireference.com/blog`. You can also find us on the twitter `@minireference` and on the facebook `fb.me/noBSguide`.

Acknowledgments

This book would not have been possible without the support and encouragement of the people around me. First and foremost of the many people to thank are my parents, who have supported my life choices, and from whom I've learned many things. Next are all my teachers. I thank my CEGEP teachers: Karnig Bedrossian from whom I learned calculus, Paul Kenton from whom I learned how to think about physics in a chill manner, and Benoit Larose who taught me that more dimensions does not mean things get more complicated. I thank Kohur Gowrisankaran, Frank Ferrie, Mourad El-Gamal, and Ioannis Psaromiligkos for their teaching during my engineering days, and Guy Moore and Zaven Altounian for teaching me advanced physics topics. Among all my teachers, I owe the most to Patrick Hayden whose teaching methods have always inspired me. From him, I learned that by defining things clearly, you can *trick* students into learning advanced topics, and even make it seem like the results are obvious! Thanks go out to all my research collaborators and friends: David Avis, Arlo Breault, Juan Pablo Di Lelle, Omar Fawzi, Adriano Ferrari, Mohamad Nizar Kezzo, Igor Khavkine, Felix Kwok, Doina Precup, Andie Sigler, and Mark M. Wilde. Thank you all for teaching me a great many things.

Preparing this book took many years and the combined efforts of many people. I want to thank Afton Lewis, Oleg Zhoglo, Georger Araujo, and Tomasz Święcicki for helping me proofread, and all the readers who reported typos and suggested clarifications. Thank you all for your comments and feedback. Above all, I want to thank my editor Sandy Gordon, who helped me substantially improve the writing in the book. Her expertise with the English language and her advice on style and content have been absolutely invaluable.

Last but not least, I want to thank all my students for their endless questions, demand for explanations, and curiosity. If I've developed any ability to explain things, I owe it to them.

Further reading

You've reached the end of this book, but you're only at the beginning of the journey of mathematical discovery. There are many things left for you to learn about. Below are some recommendations of subjects you might find interesting.

Newtonian mechanics

Newton's laws of physics provide an excellent example of the scientific approach to understanding and modelling the world around us. Using a few basic laws, it's possible to build accurate models for the motion of objects in terms of concepts like position, velocity, acceleration, momentum, and energy. It's no surprise mechanics is a foundational subject of science.

Since you know basic math, it will be easy for you to understand the laws of physics. Indeed, most physics laws are expressed as mathematical equations. If you know how to manipulate equations and solve for the unknowns in them, then you already know half of physics. If you previously avoided this subject out of fear, now would be a great time to revisit it. Learning physics is an excellent source of *knowledge buzz.*

By learning how to solve complicated physics problems, you'll also develop your analytical skills. Later, you can apply these skills to other areas of life. Even if you don't go on to study science, the expertise you develop in solving physics problems will help you tackle complicated problems in general.

[TUTORIAL] Ivan Savov. *Mechanics explained in seven pages*, 2017, see `minireference.com/static/tutorials/mech_in_7_pages.pdf`.

[BOOK] Ivan Savov. *No Bullshit Guide to Math & Physics*, Minireference Publishing, Fifth edition, 2014, 978-0-9920010-0-1.

Calculus

Calculus is the language of quantitative science. Many of the fundamental laws of physics, chemistry, biology, and economics are expressed in the language of calculus. A solid grasp of this subject is essential for anyone interested in pursuing university-level studies in science, engineering, or business.

The new knowledge you'll develop in calculus involves analyzing the properties of functions. You'll learn to describe how functions change over time (derivatives), and how to calculate the total amount of a quantity that accumulates over a time period (integrals). The language of calculus will allow you to speak precisely about the different properties of functions, and better understand their behaviour. You will learn how to calculate the slopes of functions, how to find their maximum and minimum values, how to compute their integrals, and other tasks of practical importance. In addition to its practical applications, learning calculus will also expose you to the concept of infinity. Calculus thinking requires understanding infinitely small numbers, infinitely large numbers, and computations with an infinite number of steps.

If you liked the math in this book, I encourage you to continue your journey with the **No Bullshit Guide to Math & Physics**, which will teach you the topics of mechanics, differential calculus, and integral calculus in an integrated manner, highlighting the connections between them, and discussing their applications.

[BOOK] Ivan Savov. *No Bullshit Guide to Math & Physics*, Minireference Publishing, Fifth edition, 2014, `978-0-9920010-0-1`.

The calculus book by Thompson is also worth checking out.

[BOOK] Silvanus P. Thompson. *Calculus Made Easy*, Macmillian and Co., Second edition, 1914, `http://gutenberg.org/ebooks/33283`.

You can also learn calculus from these excellent video lessons.

[VIDEOS] Grant Sanderson (3Blue1Brown), *Essence of calculus*, YouTube playlist, 2018, online: `https://bit.ly/3Blue1BrownCalc`.

[VIDEO LECTURES] Gilbert Strang. *Highlights of Calculus*, MIT OpenCourseWare, 2010, see `https://bit.ly/StrangCalcLectures`.

Linear algebra

Learning linear algebra will open many doors for you. You need linear algebra to understand statistics, computer graphics, machine learning, quantum mechanics, and many other areas of science and business.

Here's a little preview. Linear algebra is the study of vectors $\vec{v} \in \mathbb{R}^n$ and linear transformations $T : \mathbb{R}^n \to \mathbb{R}^m$. Linear transformations are *vector functions* that obey the linear property $T(\alpha\vec{v}_1 + \beta\vec{v}_2) = \alpha T(\vec{v}_1) + \beta T(\vec{v}_2)$. Using the standard notation for functions, we write $T(\vec{x}) = \vec{y}$ to show the linear transformation T acting on an input vector $\vec{x} \in \mathbb{R}^n$ to produce the output vector $\vec{y} \in \mathbb{R}^m$. Every linear transformation T can be *represented* as a matrix $A_T \in \mathbb{R}^{m \times n}$, which is an array of numbers with m rows and n columns. Computing $T(\vec{x})$ is equivalent to computing the matrix-vector product $A_T\vec{x}$. Because of the equivalence between linear transformations and matrices, we can also say that linear algebra is the study of vectors and matrices.

Vectors and matrices are used all over the place! If your knowledge of high-school math gave you modelling superpowers, then linear algebra is the vector-upgrade that teaches you how to build models in multiple dimensions.

[VIDEO LECTURES] Gilbert Strang. *Linear Algebra*, MIT OpenCourseWare, 2010, online: `http://bit.ly/StrangLAlectures`.

[BOOK] Ivan Savov. *No Bullshit Guide to Linear Algebra*, Minireference Publishing, Second edition, 2017, ISBN 978-0-9920010-2-5.

Probability

Probability distributions are a fundamental tool for modelling nondeterministic behaviour. A discrete random variable X is associated with a probability mass function $p_X(x) \stackrel{\text{def}}{=} \Pr(\{X = x\})$, which assigns a "probability mass" to each of the possible outcomes of the random variable X. For example, if X represents the outcome of the throw of a fair die, then the possible outcomes are $\mathcal{X} = \{1,2,3,4,5,6\}$ and the probability mass function has the values $p_X(x) = \frac{1}{6}, \forall x \in \mathcal{X}$.

Probability distributions and random variables allow us to model random processes like the roll of a die. We can't predict the exact outcome when two dice X_1 and X_2 are rolled, but we can predict the probability of different outcomes. For example, the "pair of sixes" outcome is described by the event $\{X_1 + X_2 = 12\}$. Assuming the dice are fair, this outcome has probability $\Pr(\{X_1 + X_2 = 12\}) = \frac{1}{36}$.

Probability theory is used all over the place, including in statistics, machine learning, quantum mechanics, and risk analysis.

[BOOK] Charles M. Grinstead and J. Laurie Snell. *Introduction to Probability*, AMS, 2006, `https://bit.ly/GrinsteadSnellProb`.

General mathematics

Mathematics is a hugely broad field. There are all kinds of topics to learn about; some of them are fun, some of them are useful, and some of them are totally mind expanding.

The following resources cover math topics of general interest and serve as a great overview of all areas of mathematics. I highly recommend you take a look at these for further math enlightening.

[VIDEO] A map of mathematics that shows all the subfields of mathematics and their objects of study: `https://youtu.be/OmJ-4B-mS-Y`.

[VIDEOS] Video interviews and lessons by some of the best math educators in the world: `https://youtube.com/user/numberphile`.

[BOOK] Alfred North Whitehead. *An Introduction to Mathematics*, Williams & Norgate, 1911, `www.gutenberg.org/ebooks/41568`.

General physics

If you want to learn more about physics, I highly recommend the Feynman lectures on physics. This three-tome collection covers all of undergraduate physics and explains many more advanced topics.

[BOOK] Richard P. Feynman. *The Feynman Lectures on Physics, The Definitive and Extended Edition*, Addison Wesley, 2005, ISBN 0805390456. Read online at: `http://feynmanlectures.caltech.edu`

Final words

Throughout this book, I strived to equip you with the basic math tools you'll need to make your future studies enjoyable and pain free. Remember to always take it easy. Never be afraid of math. Never take math exams too seriously. Play with math equations and try different approaches—the worst that can happen is you'll need another piece of paper. Grades don't matter. Big paycheques don't matter. Never settle for a boring job just because it pays well, and try to work only on projects you truly care about.

Along the way, I want you to be confident in your ability to handle any complicated math that life might throw at you. You have the tools to do anything you want; choose your own adventure.

Appendix A

Answers and solutions

Chapter 1 solutions

Answers to exercises

E1.1 a) $x = 3$; **b)** $x = 30$; **c)** $x = 2$; **d)** $x = -3$. **E1.2 a)** $\mathbb{Z}, \mathbb{Q}, \mathbb{R}, \mathbb{C}$; **b)** $\mathbb{C}$; **c)** $\mathbb{N}, \mathbb{Z}, \mathbb{Q}, \mathbb{R}, \mathbb{C}$; **d)** $\mathbb{Q}, \mathbb{R}, \mathbb{C}$; **e)** $\mathbb{R}, \mathbb{C}$. **E1.3 a)** 21; **b)** 0; **c)** $\frac{2}{27}$. **E1.4 a)** $\frac{5}{6}$; **b)** $\frac{13}{12} = 1\frac{1}{12}$; **c)** $\frac{31}{6} = 5\frac{1}{6}$. **E1.5 a)** $x = 2$; **b)** $x = 25$; **c)** $x = 100$. **E1.6 a)** $f^{-1}(x) = x^2$, $x = 16$. **b)** $g^{-1}(x) = -\frac{1}{2}\ln(x)$, $x = 0$.

Solutions to selected exercises

E1.4 a) To compute $\frac{1}{2} + \frac{1}{3}$, we rewrite both fractions using the common denominator 6, then compute the sum: $\frac{1}{2} + \frac{1}{3} = \frac{3}{6} + \frac{2}{6} = \frac{5}{6}$. **b)** You can use the answer from part (a), or compute the triple sum directly by setting all three fractions to a common denominator: $\frac{1}{2} + \frac{1}{3} + \frac{1}{4} = \frac{6}{12} + \frac{4}{12} + \frac{3}{12} = \frac{13}{12}$. **c)** Here we first rewrite $3\frac{1}{2}$ as $\frac{7}{2}$, then use the common denominator 6 for the computation: $\frac{7}{2} + 2 - \frac{1}{3} = \frac{21}{6} + \frac{12}{6} - \frac{2}{6} = \frac{31}{6}$.

Chapter 2 solutions

Answers to exercises

E2.1 a) $(x-1)(x-7)$; **b)** $(x+2)^2$; **c)** $(x+3)(x-3)$. **E2.2 a)** $x^2 + 2x - 15 = (x+1)^2 - 16 = 0$, which has solutions $x = 3$ and $x = -5$; **b)** $x^2 + 4x + 1 = (x+2)^2 - 3 = 0$, with solutions $x = -2 + \sqrt{3}$ and $x = -2 - \sqrt{3}$. **E2.3** $x_1 = \frac{3}{2}$ and $x_2 = -1$. **E2.4** $x = \pm\sqrt{2}$.

Chapter 3 solutions

Answers to exercises

E3.1 a) 8; **b)** $a^{-1}b^{-2}c^{-3} = \frac{1}{ab^2c^3}$; **c)** $8\alpha^2$; **d)** a^6b^{-2}. **E3.2 a)** 3; **b)** 12; **c)** $\sqrt{3}$; **d)** $|a|$. **E3.3**

a) 2π; **b)** $4 + \frac{1}{4} = 4.25$; **c)** 1; **d)** x^2. **E3.4 a)** $x = \sqrt{a}$ and $x = -\sqrt{a}$; **b)** $x = \sqrt[3]{b}$; **c)** $x = \sqrt[4]{c}$ and $x = -\sqrt[4]{c}$; **d)** $x = \sqrt[5]{d}$. Bonus points if you can also solve $x^2 = -1$. We'll get to that in Section 7.5. **E3.5** $k_e = 8.988 \times 10^9$. **E3.6 a)** $\log(2xy)$. **b)** $-\log(z)$. **c)** $\log(y)$. **d)** 3. **e)** -3. **f)** 4.

Solutions to selected exercises

E3.5 If you're using a very basic calculator, you should first compute the expression in the denominator, and then invert the fraction. Calculators that support scientific notation have an "`exp`" or "`E`" button, which allows you to enter ε_0 as `8.854e-12`. If your calculator supports expressions, you can type in the whole expression `1/(4*pi*8.854e-12)`. We report an answer with four significant digits because we started from a value of ε_0 with four significant digits of precision.

Chapter 5 solutions

Answers to exercises

E5.1 Domain: $\mathbb{R}$. Image: $[-2,2]$. Roots: $\{\ldots, -\frac{\pi}{2}, \frac{\pi}{2}, \frac{3\pi}{2}, \frac{5\pi}{2}, \ldots\}$. **E5.2 a)** $p(x)$ is even and has degree 4. **b)** $q(x)$ is odd and has degree 7. **E5.3 a)** $x = 5$ and $x = -3$; **b)** $x = 1 + \sqrt{3}$ and $x = 1 - \sqrt{3}$. **E5.4 a)** $(q \circ f)(x) = q(f(x)) = (x+5)^2$; $q(x)$ shifted five units to the left. **b)** $(f \circ q)(x) = x^2 + 5$; $q(x)$ shifted upward by five units. **c)** $(q \circ g)(x) = (x-6)^2$; $q(x)$ shifted six units to the right. **d)** $(q \circ h)(x) = 49x^2$; $q(x)$ horizontally compressed by a factor of seven. **E5.5** $A = 5$, $\lambda = 0.1$, and $\phi = \frac{\pi}{8}$. **E5.6** $f(x) = x^2 - 2x + 5$. **E5.7** $g(x) = 2\sqrt{x-3} - 2$.

Solutions to selected exercises

E5.3 a) Rewrite the equation putting all terms on the right-hand side: $0 = x^2 - 2x - 15$. We can factor this quadratic by inspection. Are there numbers a and b such that $a + b = -2$ and $ab = -15$? Yes, $a = -5$ and $b = 3$, so $0 = (x-5)(x+3)$. **b)** Rewrite the equation so all terms are on the left-hand side: $3x^2 - 6x - 6 = 0$. Nice, the cubic terms cancel! We'll use the quadratic formula to solve this equation $x = \frac{6 \pm \sqrt{(-6)^2 - 4(3)(-6)}}{6} = \frac{6 \pm 6\sqrt{3}}{6} = 1 \pm \sqrt{3}$.

Chapter 6 solutions

Answers to exercises

E6.1 $x = \sqrt{21}$. **E6.2** $V = 33.51$ and $A = 50.26$. **E6.3** Length of track $= 5C = 5\pi d = 11.47$ m. **E6.4** $x = 5\cos(45°) = 3.54$, $y = 5\sin(45°) = 3.54$; $C = 10\pi$. **E6.5 a)** $\frac{\pi}{6}$ rad; **b)** $\frac{\pi}{4}$ rad; **c)** $\frac{\pi}{3}$ rad; **d)** $\frac{3\pi}{2}$ rad. **E6.6 a)** -1; **b)** 1; **c)** 0. **E6.7 a)** 0; **b)** 1; **c)** $\frac{1}{2}$; **d)** 1. **E6.8 a)** $3.16\angle 18.43°$; **b)** $2.24\angle 243.43° = 2.24\angle{-116.57°}$; **c)** $6\angle 270° = 6\angle{-90°}$; **d)** $(8.66, 5)$; **e)** $(9.66, 2.59)$; **f)** $(-5, 8.66)$. **E6.9** $y = 2$. **E6.10** $c = \sqrt{a^2 - b^2}$. **E6.11** $y = \frac{1}{4f}x^2$.

Solutions to selected exercises

E6.1 The cosine rule tells us $x^2 = 4^2 + 5^2 - 2(4)(5)\cos(60°) = 21$. Therefore $x = \sqrt{21}$.

E6.2 The volume of the sphere with radius $r = 2$ is $V = \frac{4}{3}\pi 2^3 = 33.51$. Its surface area is $A = 4\pi 2^2 = 50.26$.

E6.5 To convert an angle measure from degrees to radians we must multiply it by the conversion ratio $\frac{\pi}{180}$.

E6.9 Substitute the formula $\sin\theta = \frac{y}{r}$ into the equation to obtain $r = \frac{2r}{y}$, which simplifies to $y = 2$. The function $r(\theta) = \frac{2}{\sin\theta}$ in polar coordinates corresponds to the line with equation $y = 2$. See `www.desmos.com/calculator/5n5zzoal2t` for the graph.

E6.10 First define the vertex $V_2 = (a, 0)$ which corresponds to the right extremity of the ellipse. Considering the definition of the ellipse at the vertex V_2, we find $r_1 + r_2 = (c + a) + (a - c) = 2a$. Next, consider the vertex $V_3 = (0, b)$ at the top of the ellipse. The distances r_1 and r_2 from V_3 to the focal points F_1 and F_2 correspond to the hypotenuse of a triangle with base c and height b: $r_1 = r_2 = \sqrt{c^2 + b^2}$. Since $r_1 + r_2 =$ const. for all points on the ellipse, we can equate the results obtained from the length calculations for point V_2 and point V_3. We find $2a = 2\sqrt{c^2 + b^2}$, which we can solve for c to obtain $c = \sqrt{a^2 - b^2}$.

E6.11 For a parabola with focal length f, the focal point is at $F = (0, f)$ and the directrix is the line with equation $y = -f$. The distance from the focal point to an arbitrary point on the parabola is given by

$$r = d(P, F) = d\big((x, y), (0, f)\big) = \sqrt{x^2 + (y - f)^2}.$$

The point closest to the point P that lies on the directrix is $D = (x, -f)$—the point directly below the point $P = (x, y)$. The distance from P to D is

$$\ell = d(P, D) = d\big((x, y) - (x, -f)\big) = \sqrt{(y + f)^2}.$$

The geometric definition of the parabola tells us $r = \ell$ for all points on the parabola, so we can equate the two quantities: $\sqrt{x^2 + (y - f)^2} = \sqrt{(y + f)^2}$. After squaring both sides and isolating y in the resulting expression we obtain the final equation $y = \frac{1}{4f}x^2$.

Chapter 7 solutions

Answers to exercises

E7.1 a) $(4, 0)$. **b)** $(-2, -3)$. **c)** $(7, 3)$. **E7.2 a)** $\vec{v}_1 = (5\sqrt{3}, 5) = (8.66, 5)$. **b)** $\vec{v}_2 = (0, -12)$. **c)** $\vec{v}_3 = (-2.95, 0.52)$. **E7.3 a)** $\vec{u}_1 = 4\angle 0°$. **b)** $\vec{u}_2 = \sqrt{2}\angle 45°$. **c)** $\vec{u}_3 = \sqrt{10}\angle 108.43°$. **E7.4 a)** $(1, 1, 3)$; **b)** $(1, 1, -3)$; **c)** $(3, 3, 3)$; **d)** $\sqrt{2}$. **E7.5 a)** 5; **b)** $(-1, -1, 1)$; **c)** $(1, 1, -1)$; **d)** $(0, 0, 0)$.

Chapter 8 solutions

Answers to exercises

E8.1 a) $(1, \frac{1}{2})$. **b)** $(1, 2)$. **c)** $(-2, 2)$. **E8.2** $x = 2$, $y = 3$. **E8.3** $x = 5$, $y = 6$, and $z = -3$. **E8.4** $p = 7$ and $q = 3$. **E8.5 a)** \$53 974.14; **b)** \$59 209.77; **c)** \$65 948.79. **E8.6** \$32 563.11. **E8.7 a)** $\{2, 4, 6, 7\}$; **b)** $\{1, 2, 3, 4, 5, 6\}$; **c)** $\{1, 3, 5\}$; **d)** $\varnothing$; **e)** $\{1, 2, 3, 4, 5, 6, 7\}$; **f)** $\{7\}$; **g)** $\{2, 4, 6, 7\}$; **h)** $\varnothing$. **E8.8 a)** $(-\infty, \frac{3}{2})$; **b)** $(-\infty, -5]$; **c)** $(-1, 4)$; **d)** $(4, \infty)$; **e)** $[\frac{14}{3}, \infty)$; **f)** $(-\infty, -4] \cup [2, \infty)$.

Solutions to selected exercises

E8.1 See `https://www.desmos.com/calculator/ocedywekcl` for plots.

E8.5 **a)** Since the compounding is performed monthly, we first calculate the monthly interest rate: $r = \frac{3\%}{12} = 0.25\% = 0.0025$. The sum Jack owes after 10 years is $\$40\,000(1.0025)^{120} = \$53\,974.14$. **b)** The calculation that uses the effective annual interest rate is more direct: $\$40\,000(1.04)^{10} = \$59\,209.77$. **c)** When compounding infinitely often at a nominal annual interest rate of 5%, the amount owed will grow by $\exp(\frac{5}{100}) = 1.051271$ each year. After 10 years Jack will owe $\$40\,000(1.051271)^{10} = \$65\,948.79$.

E8.6 Since there are two different interest rates in effect, we must perform two separate calculations. At the end of the first five years, Kate owes $\$20\,000(1.06)^5 = \$26\,764.51$. For the remaining five years, the interest changes to 4%, so the sum Kate owes after 10 years is $\$26\,764.51(1.04)^5 = \$32\,563.11$.

E8.8 **a)** Dividing both sides of the inequality by two gives $x < \frac{3}{2}$. **b)** Divide both sides by negative four to obtain $x \leqslant -5$. Note the "$\geqslant$" changed to "$\leqslant$" since we divided by a negative number. **c)** If the absolute value of $(2x - 3)$ is less than five, then $(2x - 3)$ must lie in the interval $(-5, 5)$. We can therefore rewrite the inequality as $-5 < 2x - 3 < 5$, then add three to both sides to obtain $-2 < 2x < 8$, and divide by two to obtain the final answer $-1 < x < 4$. **d)** Let's collect all the x-terms on the right and all the constants on the left: $8 < 2x$, which leads to $4 < x$. **e)** To simplify, add two to both sides of the inequality to obtain $\frac{1}{2}x \geqslant \frac{1}{3} + 2$. You remember how to add fractions right? We have $\frac{1}{3} + 2 = \frac{1}{3} + \frac{6}{3} = \frac{7}{3}$, and therefore $\frac{1}{2}x \geqslant \frac{7}{3}$. Multiply both sides by two to obtain $x \geqslant \frac{14}{3}$. **f)** The first step is to get rid of the square by taking the square root operation on both sides: $\sqrt{(x+1)^2} \geqslant \sqrt{9}$. Recall that $\sqrt{x^2} = |x|$, so we have $|x + 1| \geqslant 3$. There are two ways for the absolute value of $(x + 1)$ to be greater than three. Either $x + 1 \geqslant 3$ or $x + 1 \leqslant -3$. We subtract one in each of these inequalities to find $x \geqslant 2$ or $x \leqslant -4$. The solution to this inequality is the union of these two intervals.

Chapter 9 solutions

Answers to problems

P9.1 $x = \pm 4$. **P9.2** $x = A\cos(\omega t + \phi)$. **P9.3** $x = \frac{ab}{a+b}$. **P9.4** **a)** 2.2795. **b)** 1024. **c)** -8.373. **d)** 11. **P9.5** **a)** $\frac{3}{4}$. **b)** $\frac{-141}{35}$. **c)** $3\frac{23}{32}$. **P9.6** **a)** c. **b)** 1. **c)** $\frac{9|a|}{|b|}$. **d)** a. **e)** $\frac{b}{ac}$. **f)** $x^2 + ab$. **P9.7** **a)** $x^2 + (a - b)x - ab$. **b)** $2x^2 - 7x - 15$. **c)** $10x^2 + 31x - 14$. **P9.8** **a)** $(x - 4)(x + 2)$. **b)** $3x(x - 3)(x + 3)$. **c)** $(x + 3)(6x - 7)$. **P9.9** **a)** $(x - 2)^2 + 3$. **b)** $2(x + 3)^2 + 4$. **c)** $6(x + \frac{11}{12})^2 - \frac{625}{24}$. **P9.10** \$0.05. **P9.11** 13 people, 30 animals. **P9.12** 5 years later. **P9.13** girl = 80 nuts, boy = 40 nuts. **P9.14** Alice is 15. **P9.15** 18 days. **P9.16** After 2 hours. **P9.18** $\varphi = \frac{1+\sqrt{5}}{2}$. **P9.19** $x = \frac{-5 \pm \sqrt{41}}{2}$. **P9.20** **a)** $x = \sqrt[3]{2}$. **b)** $x = (\frac{\pi}{2} + 2\pi n)$ for $n \in \mathbb{Z}$. **P9.21** No real solutions if $0 < m < 8$. **P9.22** **a)** e^z. **b)** $\frac{x^3y^{15}}{z^3}$. **c)** $\frac{1}{4x^4}$. **d)** $\frac{1}{4}$. **e)** -3. **f)** $\ln(x + 1)$. **P9.23** $\epsilon = 1.110 \times 10^{-16}$; $n = 15.95$ in decimal. **P9.24** **a)** $x \in (4, \infty)$. **b)** $x \in [3, 6]$. **c)** $x \in (-\infty, -1] \cup [\frac{1}{2}, \infty)$. **P9.25** For $n > 250$, Algorithm Q is faster. **P9.26** 10 cm. **P9.27** 22.52 in. **P9.28** $h = \sqrt{3.33^2 - 1.44^2} = 3\,\text{m}$. **P9.29** The opposite side has length 1. **P9.30** $x = \sqrt{3}$, $y = 1$, and $z = 2$. **P9.31** $d = \frac{1800\tan 20° - 800\tan 25°}{\tan 25° - \tan 20°}$, $h = 1658.46\,\text{m}$. **P9.32** $x = \frac{2000}{\tan 24°}$. **P9.33** $x = \tan\theta\sqrt{a^2 + b^2 + c^2}$. **P9.34** $a = \sqrt{3}$, $A_\triangle = \frac{3\sqrt{3}}{4}$. **P9.35** $\sin^2\theta\cos^2\theta = \frac{1-\cos 4\theta}{8}$. **P9.36** $P_{\circ} = 16\tan(22.5°)$, $A_{\circ} = 8\tan(22.5°)$. **P9.37** $c = \frac{a\sin 75°}{\sin 41°} \approx 14.7$. **P9.38**

a) $h = a\sin\theta$. **b)** $A = \frac{1}{2}ba\sin\theta$. **c)** $c = \sqrt{a^2+b^2-2ab\cos(180^\circ-\theta)}$. **P9.39** $B = 44.8^\circ$, $C = 110.2^\circ$. $c = \frac{a\sin 110.2^\circ}{\sin 25^\circ} \approx 39.97$. **P9.40** $v = 742.92$ km/h. **P9.41** 1.06 cm. **P9.42** $x = 9.55$. **P9.43** $\frac{1}{2}(\pi 4^2 - \pi 2^2) = 18.85\text{ cm}^2$. **P9.44** $\ell_{\text{rope}} = 8.42\,\text{m}$. **P9.45** $A_{\text{rect}} = 5c + 10$. **P9.46** $V_{\text{box}} = 1.639\,\text{L}$. **P9.47** $\frac{R}{r} = \frac{1-\sin 15^\circ}{\sin 15^\circ} = 2.8637$. **P9.48** 7 cm. **P9.49** $V = 300\,000\,\text{L}$. **P9.50** 315 000 L. **P9.51** 4000 L. **P9.52** $d = \frac{1}{2}(35 - 5\sqrt{21})$. **P9.53** A rope of length $\sqrt{2}\ell$. **P9.54** 20 L of water. **P9.55** $h = 7.84$ inches. **P9.56** $1 + 2 + \cdots + 100 = 50 \times 101 = 5050$. **P9.57 a)** $\vec{u}_1 = 5\angle 90^\circ$. **b)** $\vec{u}_2 = \sqrt{5}\angle 63.4^\circ$. **c)** $\vec{u}_3 = \sqrt{5}\angle 243.4^\circ$ or $\sqrt{5}\angle{-116.6^\circ}$. **P9.58 a)** $\vec{v}_1 = (17.32, 10)$. **b)** $\vec{v}_2 = (0, -10)$. **c)** $\vec{v}_3 = (-4.33, 2.5)$. **P9.59 a)** $\vec{w}_1 = 9.06\hat{\imath} + 4.23\hat{\jmath}$. **b)** $\vec{w}_2 = -7\hat{\jmath}$. **c)** $\vec{w}_3 = 3\hat{\imath} - 2\hat{\jmath} + 3\hat{k}$. **P9.60 a)** $(3,4)$. **b)** $(0,1)$. **c)** $(7.33, 6.5)$. **P9.61** $Q = (5.73, 4)$. **P9.62 a)** 6. **b)** 0. **c)** -3. **d)** $(-2,1,1)$. **e)** $(3,-3,0)$. **f)** $(7,-5,1)$. **P9.63 a)** $(2,3,3,7,8)$. **b)** $(0,-1,-3,-1,-2)$. **c)** 30. **P9.64** $(-\frac{2}{3}, \frac{1}{3}, \frac{2}{3})$ or $(\frac{2}{3}, -\frac{1}{3}, -\frac{2}{3})$. **P9.65** $(12, -4, -12)$. **P9.66 a)** $2i$. **b)** $\frac{1}{4}(5+i)$. **c)** $2+i$. **P9.67 a)** $x = \pm 2i$. **b)** $x = -16$. **c)** $x = -1 - i$ and $x = -1 + i$. **d)** $x = i$, $x = -i$, $x = \sqrt{3}i$, and $x = -\sqrt{3}i$. **P9.68 a)** $\sqrt{5}$. **b)** $\frac{1}{2}(-3+i)$. **c)** $-5 - 5i$. **P9.69** $t = 52$ weeks. **P9.70** The tractor's trajectory is a half-circle. The total distance travelled is 3.14 km. **P9.71** $x = -2$ and $y = 2$. **P9.72** $x = 1$, $y = 2$, and $z = 3$. **P9.73** \$112. **P9.74** 20%. **P9.75** \$16501.93. **P9.77** 0.14 s. **P9.78** $\tau = 34.625\,\text{min}$, 159.45 min. **P9.79** $V(0.01) = 15.58$ volts. $V(0.1) = 1.642$ volts. **P9.83** $A_1(x) = 3x$ and $A_2(x) = \frac{1}{2}x^2$.

Solutions to selected problems

P9.5 For **c)**, $1\frac{3}{4} + 1\frac{31}{32} = \frac{7}{4} + \frac{63}{32} = \frac{56}{32} + \frac{63}{32} = \frac{119}{32} = 3\frac{23}{32}$.

P9.9 The solutions for **a)** and **b)** are fairly straightforward. To solve **c)**, we first factor out 6 from the first two terms to obtain $6(x^2 + \frac{11}{6}x) - 21$. Next we choose half of the coefficient of the linear term to go inside the square and add the appropriate correction to maintain equality: $6[x^2 + \frac{11}{6}x] - 21 = 6[(x + \frac{11}{12})^2 - (\frac{11}{12})^2] - 21$. After expanding the rectangular brackets and simplifying, we obtain the final expression: $6(x + \frac{11}{12})^2 - \frac{625}{24}$.

P9.11 Let p denote the number of people and a denote the number of animals. We are told $p + a = 43$ and $a = p + 17$. Substituting the second equation into the first, we find $p + (p + 17) = 43$, which is equivalent to $2p = 26$ or $p = 13$. There are 13 figures of people and 30 figures of animals.

P9.12 We must solve for x in $35 + x = 4(5 + x)$. We obtain $35 + x = 20 + 4x$, then $15 = 3x$, so $x = 5$.

P9.14 Let A be Alice's age and B be Bob's age. We're told $A = B + 5$ and $A + B = 25$. Substituting the first equation into the second we find $(B + 5) + B = 25$, which is the same as $2B = 20$, so Bob is 10 years old. Alice is 15 years old.

P9.15 The first shop can bind $4500/30 = 150$ books per day. The second shop can bind $4500/45 = 100$ books per day. The combined production capacity rate is $150 + 100 = 250$ books per day. It will take $4500/250 = 18$ days to bind the books when the two shops work in parallel.

P9.16 Let t_{m} denote the time when the two planes meet, as measured from the moment the second plane departs. Since it left one hour earlier, the slower plane will have travelled a distance $600(t_{\text{m}} + 1)$ km when they meet. The faster plane will have travelled the distance $900t_{\text{m}}$ km when they meet. Combining the two expressions we find $600(t_{\text{m}} + 1) = 900t_{\text{m}}$. The time when the planes meet is $t_{\text{m}} = 2$ hours after the departure of the second plane.

P9.17 This is a funny nonsensical problem that showed up on a school exam. I'm just checking to make sure you're still here.

P9.21 Using the quadratic formula, we find $x = \frac{m \pm \sqrt{m^2 - 8m}}{4}$. If $m^2 - 8m \geqslant 0$, the solutions are real. If $m^2 - 8m < 0$, the solutions will be complex numbers. Factoring the expressions and plugging in some numbers, we observe that $m^2 - 8m = m(m - 8) < 0$ for all $m \in (0, 8)$.

P9.23 See `bit.ly/float64prec` for the calculations.

P9.24 For **c)**, complete the square on the left side: $2x^2 + x = 2(x + \frac{1}{4})^2 - \frac{1}{8}$. The inequality $2x^2 + x \geqslant 1$ can be rewritten as $2(x + \frac{1}{4})^2 \geqslant 1 + \frac{1}{8} = \frac{9}{8}$. Dividing by 2 on both sides gives $(x + \frac{1}{4})^2 \geqslant \frac{9}{16}$. Taking the square root produces $|x + \frac{1}{4}| \geqslant \frac{3}{4}$, which is satisfied if $x \geqslant \frac{1}{2}$ or if $x \leqslant -1$. Use the union operation to combine the two parts of the solution interval.

P9.25 The running time of Algorithm Q grows linearly with the size of the problem, whereas Algorithm P's running time grows quadratically. To find the size of the problem when the algorithms take the same time, we solve $P(n) = Q(n)$, which is $0.002n^2 = 0.5n$. The solution is $n = 250$. For $n > 250$, the linear-time algorithm (Algorithm Q) will take less time.

P9.29 Solve for b in Pythagoras' formula $c^2 = a^2 + b^2$ with $c = \varphi$, and $a = \sqrt{\varphi}$. The triangle with sides 1, $\sqrt{\varphi}$, and φ is called Kepler's triangle.

P9.30 Use Pythagoras' theorem to find x. Then use $\cos(30°) = \frac{\sqrt{3}}{2} = \frac{x}{z}$ to find z. Finally use $\sin(30°) = \frac{1}{2} = \frac{y}{z}$ to find y.

P9.31 Observe the two right-angle triangles drawn in Figure 9.1. From the triangle with angle 25° we know $\tan 25° = \frac{h}{800+d}$. From the triangle with angle 20° we know $\tan 20° = \frac{h}{1800+d}$. We isolate h in both equations and eliminate h by equating $(1800 + d)\tan 25° = \tan 20°(800 + d)$. Solving for d we find $d = \frac{1800 \tan 20° - 800 \tan 25°}{\tan 25° - \tan 20°} = 2756.57\,\text{m}$. Finally we use $\tan 25° = \frac{h}{800+d}$ again to obtain $h = \tan 25°(800 + d) = 1658.46\,\text{m}$.

P9.32 Consider the right-angle triangle with base x and opposite side 2000. Looking at the diagram we see that $\theta = 24°$. We can then use the relation $\tan 24° = \frac{2000}{x}$ and solve for x.

P9.34 The internal angles of an equilateral triangle are all 60°. Draw three radial lines that connect the centre of the circle to each vertex of the triangle. The equilateral triangle is split into three obtuse triangles with angle measures 30°, 30°, and 120°. Split each of these obtuse sub-triangles down the middle to obtain six right-angle triangles with hypotenuse 1. The side of the equilateral triangle is equal to two times the base of the right-angle triangles $a = 2\cos(30°) = \sqrt{3}$. To find the area, we use $A_\triangle = \frac{1}{2}ah$, where $h = 1 + \sin(30°)$.

P9.35 We know $\sin^2(\theta) = \frac{1}{2}(1 - \cos(2\theta))$ and $\cos^2(\theta) = \frac{1}{2}(1 + \cos(2\theta))$, so their product is $\frac{1}{4}(1 - \cos(2\theta)\cos(2\theta))$, and $\cos(2\theta)\cos(2\theta) = \cos^2(2\theta)$. Using the power-reduction formula on the term $\cos^2(2\theta)$ gives $\sin^2\theta\cos^2\theta = \frac{1}{4}\left(1 - \frac{1}{2}(1 + \cos(4\theta))\right)$.

P9.36 Split the octagon into eight isosceles triangles. The height of each triangle will be 1, and its angle measure at the centre will be $\frac{360°}{8} = 45°$. Split each of these triangles into two halves down the middle. The octagon is now split into 16 similar right-angle triangles with angle measure 22.5° at the centre. In a right-angle triangle with angle 22.5° and adjacent side 1, what is the length of the opposite side? The opposite side of each of the 16 triangles is $\frac{b}{2} = \tan(22.5°)$, so the perimeter of the octagon is $P_{\octagon} = 16\tan(22.5°)$. In general, if a unit circle is inscribed inside an n-sided regular polygon, the perimeter of the polygon is $P_n = 2n\tan\left(\frac{360°}{2n}\right)$. To find the area of the octagon, we use the formula $A_\triangle = \frac{1}{2}bh$, with $b = 2\tan(22.5°)$ and $h = 1$ to find the area of each isosceles triangle. The area of the octagon is $A_{\circ} = 8 \cdot \frac{1}{2}(2\tan(22.5°))(1) = 8\tan(22.5°)$. For an n-sided regular polygon the area formula is $A_n = n\tan\left(\frac{360°}{2n}\right)$. Bonus points if you can tell me what happens to the formulas for P_n and A_n as n goes to infinity (see `bit.ly/1jGU1Kz`).

P9.40 Initially the horizontal distance between the observer and the plane is $d_1 = \frac{2000}{\tan 30°}$ m. After 10 seconds, the distance is $d_2 = \frac{2000}{\tan 55°}$ m. Velocity is change in distance divided by the time $v = \frac{d_1 - d_2}{10} = 206.36$ m/s. To convert m/s into km/h, we must multiply by the appropriate conversion factors: $206.36\,\text{m/s} \times \frac{1\,\text{km}}{1000\,\text{m}} \times \frac{3600\,\text{s}}{1\,\text{h}} = 742.92$ km/h.

P9.41 The volume of the water stays constant and is equal to $1000\,\text{cm}^3$. Initially the height of the water h_1 can be obtained from the formula for the volume of a cylinder $1000\,\text{cm}^3 = h_1\pi(8.5\,\text{cm})^2$, so $h_1 = 4.41\,\text{cm}$. After the bottle is inserted, the water has the shape of a cylinder with a cylindrical part missing. The volume of water is $1000\,\text{cm}^3 = h_2\left(\pi(8.5\,\text{cm})^2 - \pi(3.75\,\text{cm})^2\right)$. We find $h_2 = 5.47\,\text{cm}$. The change in height is $h_2 - h_1 = 5.47 - 4.41 = 1.06\,\text{cm}$.

P9.42 Using the law of cosines for the angles α_1 and α_2, we obtain the equations $7^2 = 8^2 + 12^2 - 2(8)(12)\cos\alpha_1$ and $11^2 = 4^2 + 12^2 - 2(4)(12)\cos\alpha_2$ from which we find $\alpha_1 = 34.09°$ and $\alpha_2 = 66.03°$. In the last step we use the law of cosines again to obtain $x^2 = 8^2 + 4^2 - 2(8)(4)\cos(34.09° + 66.03°)$.

P9.44 The length of the horizontal part of the rope is $\ell_h = 4\sin 40$. The circular portion of the rope that hugs the pulley has length $\frac{1}{4}$ of the circumference of a circle with radius $r = 50\,\text{cm} = 0.5\,\text{m}$. Using the formula $C = 2\pi r$, we find $\ell_c = \frac{1}{4}(2\pi(0.5)) = \frac{\pi}{4}$. The vertical part of the rope has length $\ell_v = 4\cos 40 + 2$. The total length of rope is $\ell_h + \ell_c + \ell_v = 8.42\,\text{m}$.

P9.45 The rectangle's area is equal to its length times its height $A_{\text{rect}} = \ell h$.

P9.46 The box's volume is $V = w \times h \times \ell = 10.5 \times 7 \times 22.3 = 1639\,\text{cm}^3 = 1.639\,\text{L}$.

P9.47 The base of this triangle has length $2r$ and each side has length $R + r$. If you split this triangle through the middle, each half is a right triangle with an angle at the centre $\frac{360°}{24} = 15°$, hypotenuse $R + r$, and opposite side r. We therefore have $\sin 15° = \frac{r}{R+r}$. After rearranging this equation, we find $\frac{R}{r} = \frac{1-\sin 15°}{\sin 15°} = 2.8637$.

P9.50 The tank's total capacity is $15 \times 6 \times 5 = 450\,\text{m}^3$. If 30% of its capacity is spent, then 70% of the capacity remains: $315\,\text{m}^3$. Knowing that $1\,\text{m}^3 = 1000\,\text{L}$, we find there are 315 000 L in the tank.

P9.51 The first tank contains $\frac{1}{4} \times 4000 = 1000$ L. The second tank contains three times more water, so 3000 L. The total is 4000 L.

P9.52 Let's define w and h to be the width and the height of the hole. Define d to be the distance from the hole to the sides of the lid. The statement of the problem dictates the following three equations must be satisfied: $w + 2d = 40$, $h + 2d = 30$, and $wh = 500$. After some manipulations, we find $w = 5(1 + \sqrt{21})$, $h = 5(\sqrt{21} - 1)$ and $d = \frac{1}{2}(35 - 5\sqrt{21})$.

P9.53 The amount of wood in a pack of wood is proportional to the area of a circle $A = \pi r^2$. The circumference of this circle is equal to the length of the rope $C = \ell$. Note the circumference is proportional to the radius $C = 2\pi r$. If we want double the area, we need the circle to have radius $\sqrt{2}r$, which means the circumference needs to be $\sqrt{2}$ times larger. If we want a pack with double the wood, we need to use a rope of length $\sqrt{2}\ell$.

P9.54 In 10 L of a 60% acid solution there are 6 L of acid and 4 L of water. A 20% acid solution will contain four times as much water as it contains acid, so 6 L acid and 24 L water. Since the 10 L we start from already contains 4 L of water, we must add 20 L.

P9.55 The document must have a 768/1004 aspect ratio, so its height must be $6 \times \frac{1004}{768} = 7.84375$ inches.
P9.56 If we rewrite $1+2+3+\cdots+98+99+100$ by pairing numbers, we obtain the sum $(1+100)+(2+99)+(3+98)+\cdots$. This list has 50 terms and each term has the value 101. Therefore $1+2+3+\cdots+100 = 50 \times 101 = 5050$.
P9.64 See `bit.ly/1c0a8yo` for calculations.
P9.65 Any multiple of the vector $\vec{u}_1 \times \vec{u}_2 = (-3,1,3)$ is perpendicular to both $\vec{u}_1$ and $\vec{u}_2$. We must find a multiplier $t \in \mathbb{R}$ such that $t(-3,1,3)\cdot(1,1,0) = 8$. Computing the dot product we find $-3t + t = 8$, so $t = -4$. The vector we're looking for is $(12,-4,-12)$. See `bit.ly/1nmYH8T` for calculations.
P9.69 We want the final state of the project to be 100% real: $p_f = 100$. Since we start from $p_o = 100i$, the rotation required is $e^{-i\alpha h(t)} = e^{-i\frac{\pi}{2}}$, which means $\alpha h(t) = \frac{\pi}{2}$. We can rewrite this equation as $h(t) = 0.2t^2 = \frac{\pi}{2\alpha}$ and solving for t we find $t = \sqrt{\frac{\pi}{2(0.002904)(0.2)}} = 52$ weeks.
P9.70 The direction of the tractor changes constantly throughout the day, and the overall trajectory has the shape of a half-circle. The total distance travelled by the tractor is equal to half the circumference of a circle of radius R. Since it took the tractor six hours of movement at $v = 0.524$ km/h to travel half the circumference of the circle, we have $\frac{1}{2}C = \pi R = v(t_f - t_i) = 0.524(6)$, from which we find $R = 1$ km. The total distance travelled by the tractor is $\pi R = 3.14$ km.
P9.75 An nAPR of 12% means the monthly interest rate is $\frac{12\%}{12} = 1\%$. After 10 years you'll owe $\$5000(1.01)^{120} = \$16501.93$. Yikes!
P9.76 The graphs of the functions are shown in Figure A.1. Observe that $f(x)$ decreases to 37% of its initial value when $x = 2$. The increasing exponential $g(x)$ reaches 63% of its maximum value at $x = 2$.

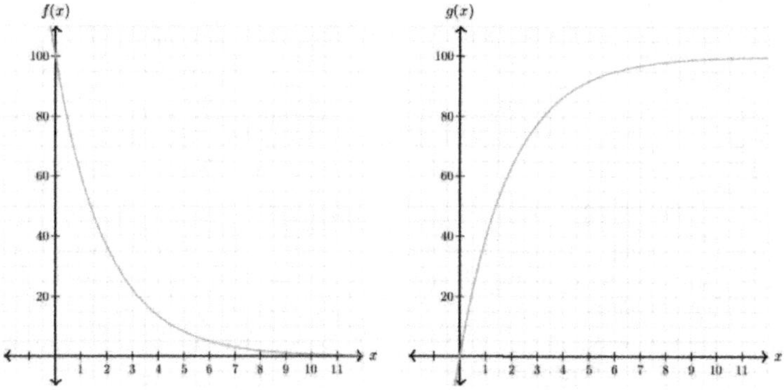

(a) The graph of $f(x)$. (b) The graph of $g(x)$.

Figure A.1: The graphs of the two functions from **P9.76**.

P9.77 We're looking for the time t when $Q(t)/Q_o = \frac{1}{2}$, which is the same as $e^{-5t} = 0.5$. Take logarithms of both sides to find $-5t = \ln(0.5)$ and solve for t to get $t = 0.14$ s.
P9.78 We're told $T(24)/T_o = \frac{1}{2} = e^{-24/\tau}$, which we can rewrite as $\ln(\frac{1}{2}) = -24/\tau$. Solving for τ, we find $\tau = \frac{24}{\ln 2} = 34.625$ min. To find the time the body takes to reach 1% of its initial temperature, we must solve for t in $T(t)/T_o = 0.01 = e^{-t/34.625}$. We find $t = 159.45$ min.
P9.80 There exists at least one banker who is not a crook. Another way of saying the same thing is "not all bankers are crooks"—just *most* of them.
P9.81 Everyone steering the ship at Monsanto ought to burn in hell, forever.
P9.82 **a)** Investors with money but without connections. **b)** Investors with connections but no money. **c)** Investors with both money and connections.

Appendix B

Notation

This appendix contains a summary of the notation used in this book.

Math notation

Expression	Read as	Used to denote
a, b, x, y		variables
$=$	is equal to	expressions that have the same value
$\stackrel{\text{def}}{=}$	is defined as	a new variable definition
$a + b$	a plus b	the combined lengths of a and b
$a - b$	a minus b	the difference in lengths between a and b
$a \times b = ab$	a times b	the area of a rectangle
$a^2 = aa$	a squared	the area of a square of side length a
$a^3 = aaa$	a cubed	the volume of a cube of side length a
a^n	a to the n	a multiplied by itself n times
$\sqrt{a} = a^{\frac{1}{2}}$	square root of a	the side length of a square of area a
$\sqrt[3]{a} = a^{\frac{1}{3}}$	cube root of a	the side length of a cube with volume a
$a/b = \frac{a}{b}$	a divided by b	a parts of a whole split into b parts
$a^{-1} = \frac{1}{a}$	one over a	division by a
$f(x)$	f of x	the function f applied to input x
f^{-1}	f inverse	the inverse function of $f(x)$
$f \circ g$	f compose g	function composition; $f \circ g(x) = f(g(x))$
e^x	e to the x	the exponential function base e
$\ln(x)$	natural log of x	the logarithm base e
a^x	a to the x	the exponential function base a
$\log_a(x)$	log base a of x	the logarithm base a
θ, ϕ	*theta, phi*	angles
sin, cos, tan	sin, cos, tan	trigonometric ratios
%	percent	proportions of a total; $a\% = \frac{a}{100}$

Set notation

You don't need a lot of fancy notation to understand mathematics. It really helps, though, if you know a little bit of set notation.

Symbol	Read as	Denotes
$\{\ldots\}$	the set ...	define a sets
$\mid$	such that	describe or restrict the elements of a set
$\mathbb{N}$	the naturals	the set $\mathbb{N} \stackrel{\text{def}}{=} \{0,1,2,\ldots\}$. Note $\mathbb{N}^* \stackrel{\text{def}}{=} \mathbb{N}\backslash\{0\}$.
$\mathbb{Z}$	the integers	the set $\mathbb{Z} \stackrel{\text{def}}{=} \{\ldots,-2,-1,0,1,2,3,\ldots\}$
$\mathbb{Q}$	the rationals	the set of fractions of integers
$\mathbb{R}$	the reals	the set of real numbers
$\mathbb{C}$		the set of complex numbers
$\subset$	subset	one set strictly contained in another
$\subseteq$	subset or equal	containment or equality
$\cup$	union	the combined elements from two sets
$\cap$	intersection	the elements two sets have in common
$S\backslash T$	S set minus T	the elements of S that are not in T
$a \in S$	a in S	a is an element of set S
$a \notin S$	a not in S	a is not an element of set S
$\forall x$	for all x	a statement that holds for all x
$\exists x$	there exists x	an existence statement
$\nexists x$	there doesn't exist x	a non-existence statement

An example of a condensed math statement that uses set notation is "$\nexists m, n \in \mathbb{Z}$ such that $\frac{m}{n} = \sqrt{2}$," which reads "there don't exist integers m and n whose ratio equals $\sqrt{2}$." Since we identify the set of fractions of integers with the rationals, this statement is equivalent to the shorter "$\sqrt{2} \notin \mathbb{Q}$," which reads "$\sqrt{2}$ is irrational."

Complex numbers notation

Expression	Denotes
$\mathbb{C}$	the set of complex numbers $\mathbb{C} \stackrel{\text{def}}{=} \{a+bi \mid a,b \in \mathbb{R}\}$
i	the unit imaginary number $i \stackrel{\text{def}}{=} \sqrt{-1}$ or $i^2 = -1$
$\text{Re}\{z\} = a$	real part of $z = a + bi$
$\text{Im}\{z\} = b$	imaginary part of $z = a + bi$
$\lvert z\rvert\angle\varphi_z$	polar representation of $z = \lvert z\rvert\cos\varphi_z + i\lvert z\rvert\sin\varphi_z$
$\lvert z\rvert = \sqrt{a^2+b^2}$	magnitude of $z = a + bi$
$\varphi_z = \tan^{-1}(b/a)$	phase or argument of $z = a + bi$
$\bar{z} = a - bi$	complex conjugate of $z = a + bi$

Vectors notation

Expression	Denotes
$\mathbb{R}^n$	the set of n-dimensional vectors
$\vec{v}$	a vector
(v_x, v_y)	vector in component notation
$v_x\hat{\imath} + v_y\hat{\jmath}$	vector in unit vector notation
$\\|\vec{v}\\|\angle\theta$	vector in length-and-direction notation
$\\|\vec{v}\\|$	length of the vector $\vec{v}$
θ	angle the vector $\vec{v}$ makes with the x-axis
$\hat{v} \stackrel{\text{def}}{=} \frac{\vec{v}}{\\|\vec{v}\\|}$	unit vector in the same direction as $\vec{v}$
$\vec{u} \cdot \vec{v}$	dot product of the vectors $\vec{u}$ and $\vec{v}$
$\vec{u} \times \vec{v}$	cross product of the vectors $\vec{u}$ and $\vec{v}$

Appendix C

SymPy tutorial

Computers can be very useful for dealing with complicated math expressions or when slogging through tedious calculations. Throughout this book we used `SymPy` to illustrate several math concepts. We'll now review all the math tools available through the `SymPy` command line. Don't worry if you're not a computer person; we'll only discuss concepts we covered in the book, and the computer commands we'll learn are very similar to the math operations you're already familiar with. This section also serves as a final review of the material covered in the book.

Introduction

You can use a computer algebra system (CAS) to manipulate complicated math expressions and solve any equation.

All computer algebra systems offer essentially the same functionality, so it doesn't matter which system you use: there are free systems like `SymPy`, `Magma`, or `Octave`, and commercial systems like `Maple`, `MATLAB`, and `Mathematica`. This tutorial is an introduction to `SymPy`, which is a *symbolic* computer algebra system written in the programming language Python. In a symbolic CAS, numbers and operations are represented symbolically, so the answers obtained are exact. For example, the number $\sqrt{2}$ is represented in `SymPy` as the object `Pow(2,1/2)`, whereas in *numerical* computer algebra systems like `Octave`, the number $\sqrt{2}$ is represented as the approximation 1.41421356237310 (a `float`). For most purposes the approximation is okay, but sometimes approximations can lead to problems: `float(sqrt(2))*float(sqrt(2))` = `2.00000000000000044` $\neq$ 2. Because `SymPy` uses exact representations, you'll never run into such problems: `Pow(2,1/2)*Pow(2,1/2)`= 2.

This tutorial presents many explanations as code snippets. Be sure to try the code examples on your own by typing the commands into SymPy. It's always important to verify for yourself!

Using SymPy

The easiest way to use SymPy, provided you're connected to the internet, is to visit http://live.sympy.org. You'll be presented with an interactive prompt into which you can enter your commands—right in your browser.

If you want to use SymPy on your own computer, you must first install Python and the Python package sympy. You can then open a command prompt and start a Python session using:

```
you@host> python
Python X.Y.Z
[GCC a.b.c (Build Info)] on platform
Type "help", "copyright", or "license" for more information.
>>>
```

The >>> prompt indicates you're in the Python shell which accepts Python commands. Type the following in the Python shell:

```
>>> from sympy import *
>>>
```

The command `from sympy import *` imports all the SymPy functions into the current namespace. All SymPy functions are now available to you. To exit the python shell press CTRL+D.

For an even better experience, you can try `jupyter notebook`, which is a web interface for accessing the Python shell. Search the web for "jupyter notebook" and follow the installation instructions specific to your operating system. It's totally worth it!

Each section in this appendix begins with a python `import` statement for the functions used in that section. If you use the statement `from sympy import *` in the beginning of your code, you don't need to run these individual import statements, but I've included them so you'll know which SymPy vocabulary is covered in each section.

Fundamentals of mathematics

Let's begin by learning about the basic SymPy objects and the operations we can carry out on them. We'll learn the SymPy equivalents of many math verbs like: "to solve" (an equation), "to expand" (an expression), "to factor" (a polynomial).

Numbers

```
>>> from sympy import  sympify, S, evalf, N
```

In Python, there are two types of number objects: ints and floats.

```
>>> 3
3                               # an int
>>> 3.0
3.0                             # a float
```

Integer objects in Python are a faithful representation of the set of integers $\mathbb{Z} = \{\ldots, -2, -1, 0, 1, 2, \ldots\}$. Floating point numbers are approximate representations of the reals $\mathbb{R}$. A floating point number has 16 decimals of precision.

Special care is required when specifying rational numbers if you want to get exact answers. If you try to divide two numbers, Python will compute a floating point approximation:

```
>>> 1/7
0.14285714285714285             # a float
```

The floating point number 0.14285714285714285 is an approximation of the exact number $\frac{1}{7} \in \mathbb{Q}$. The float approximation has 16 decimals while the decimal expansion of $\frac{1}{7}$ is infinitely long. To obtain an *exact* representation of $\frac{1}{7}$ you need to create a SymPy expression. You can sympify any expression using the shortcut function S():

```
>>> S('1/7')                    # = sympify('1/7')
1/7                             # = Rational(1,7)
```

Note the input to S() is specified as a text string delimited by quotes. We could have achieved the same result using S('1')/7 since a SymPy object divided by an int is a SymPy object.

Except for the tricky Python division operator, other math operators like addition +, subtraction -, and multiplication * work as you would expect. The syntax ** is used to denote exponentiation:

```
>>> 2**10                       # same as S('2^10')
1024
```

When solving math problems, it's best to work with SymPy objects, and wait to compute the numeric answer in the end. To obtain a numeric approximation of a SymPy object as a float, call its .evalf() method:

```
>>> pi
pi
>>> pi.evalf()                  # = pi.n() = N(pi)
3.14159265358979
```

The method .n() is equivalent to .evalf(). The global SymPy function N() can also be used to to compute numerical values. You can easily change the number of digits of precision of the approximation. Enter pi.n(400) to obtain an approximation of π to 400 decimals.

Symbols

```
>>> from sympy import Symbol, symbols
```

Python is a civilized language so there's no need to define variables before assigning values to them. When you write `a = 3`, you define a new name `a` and set it to the value 3. You can now use the name `a` in subsequent calculations.

Most interesting `SymPy` calculations require us to define `symbols`, which are the `SymPy` objects for representing variables and unknowns. For your convenience, when `live.sympy.org` starts, it runs the following commands automatically:

```
>>> from sympy import *
>>> x, y, z, t = symbols('x y z t')
>>> k, m, n = symbols('k m n', integer=True)
>>> f, g, h = symbols('f g h', cls=Function)
```

The first statement imports all the `SymPy` functions. The other three statements define some generic symbols `x`, `y`, `z`, and `t`, and several other symbols with special properties.

Note the difference between the following two statements:

```
>>> x + 2
x + 2                   # an Add expression
>>> p + 2
NameError: name 'p' is not defined
```

The name `x` is defined as a symbol, so `SymPy` knows that `x + 2` is an expression; but the variable `p` is not defined, so `SymPy` doesn't know what to make of `p + 2`. To use `p` in expressions, you must first define it as a symbol:

```
>>> p = Symbol('p')    # the same as p = symbols('p')
>>> p + 2
p + 2                   # = Add(Symbol('p'), Integer(2))
```

You can define a sequence of variables using the following notation:

```
>>> a0, a1, a2, a3 = symbols('a0:4')
```

You can use any name you want for a variable, but it's best if you avoid the letters `Q`,`C`,`O`,`S`,`I`,`N` and `E` because they have special uses in `SymPy`: `I` is the unit imaginary number $i \stackrel{\text{def}}{=} \sqrt{-1}$, `E` is the base of the natural logarithm, `S()` is the `sympify` function, `N()` is used to obtain numeric approximations, and `O` is used for big-`O` notation.

Expressions

```
>>> from sympy import simplify, factor, expand, collect
```

You define `SymPy` expressions by combining symbols with basic math operations and other functions:

```
>>> expr = 2*x + 3*x - sin(x) - 3*x + 42
>>> simplify(expr)                    # simplify the expression
2*x - sin(x) + 42
```

The function `simplify` can be used on any expression to simplify it. The examples below illustrate other useful `SymPy` functions that correspond to common mathematical operations on expressions:

```
>>> factor( x**2-2*x-8 )              # factor a polynomial
(x - 4)*(x + 2)
>>> expand( (x-4)*(x+2) )             # expand and expression
x**2 - 2*x - 8
>>> collect(x**2+x*b+a*x+a*b, x)      # collect like terms in x
x**2 + (a+b)*x + a*b
```

To substitute a given value into an expression, call the `.subs()` method, passing in a python dictionary object `{ key:val, ... }` with the symbol–value substitutions you want to make:

```
>>> expr  = sin(x) + cos(y)           # define an expression
>>> expr
sin(x) + cos(y)
>>> expr.subs({x:1, y:2})             # subs. x=1,y=1 in expr
sin(1) + cos(2)
>>> expr.subs({x:1, y:2}).n()         # compute numeric value
0.425324148260754
```

Note how we used `.n()` to obtain the expression's numeric value.

Solving equations

```
>>> from sympy import solve
```

The function `solve` is the main workhorse in `SymPy`. This incredibly powerful function knows how to solve all kinds of equations. In fact `solve` can solve pretty much *any* equation! When high school students learn about this function, they get really angry—why did they spend five years of their life learning to solve various equations by hand, when all along there was this `solve` thing that could do all the math for them? Don't worry, learning math is *never* a waste of time.

The function `solve` takes two arguments. Use `solve(expr,var)` to solve the equation `expr==0` for the variable `var`. You can rewrite any equation in the form `expr==0` by moving all the terms to one side of the equation; the solutions to $A(x) = B(x)$ are the same as the solutions to $A(x) - B(x) = 0$.

For example, to solve the quadratic equation $x^2 + 2x - 8 = 0$, use

```
>>> solve( x**2 + 2*x - 8, x)
[2, -4]
```

In this case the equation has two solutions so `solve` returns a list. Check that $x = 2$ and $x = -4$ satisfy the equation $x^2 + 2x - 8 = 0$.

The best part about `solve` and `SymPy` is that you can obtain symbolic answers when solving equations. Instead of solving one specific quadratic equation, we can solve all possible equations of the form $ax^2 + bx + c = 0$ using the following steps:

```
>>> a, b, c = symbols('a b c')
>>> solve( a*x**2 + b*x + c, x)
[(-b+sqrt(b**2 - 4*a*c))/(2*a), (-b-sqrt(b**2-4*a*c))/(2*a)]
```

In this case `solve` calculated the solution in terms of the symbols `a`, `b`, and `c`. You should be able to recognize the expressions in the solution—it's the quadratic formula $x_{1,2} = \frac{-b\pm\sqrt{b^2-4ac}}{2a}$.

To solve a *system of equations*, you can feed `solve` with the list of equations as the first argument, and specify the list of unknowns you want to solve for as the second argument. For example, to solve for x and y in the system of equations $x + y = 3$ and $3x - 2y = 0$, use

```
>>> solve([x + y - 3, 3*x - 2*y], [x, y])
{x: 6/5, y: 9/5}
```

The function `solve` is like a Swiss Army knife you can use to solve all kind of problems. Suppose you want to *complete the square* in the expression $x^2 - 4x + 7$, that is, you want to find constants h and k such that $x^2 - 4x + 7 = (x - h)^2 + k$. There is no special "complete the square" function in `SymPy`, but you can call `solve` on the equation $(x - h)^2 + k \; - \; (x^2 - 4x + 7) = 0$ to find the unknowns h and k:

```
>>> h, k = symbols('h k')
>>> solve( (x-h)**2 + k  - (x**2-4*x+7), [h,k] )
[(2, 3)]                                   # so h = 2 and k = 3
>>> expand( (x-2)**2+3 )                   # verify...
x**2 - 4*x + 7
```

Learn the basic `SymPy` commands and you'll never need to suffer another tedious arithmetic calculation painstakingly performed by hand again!

Rational functions

```
>>> from sympy import together, apart
```

By default, `SymPy` will not combine or split rational expressions. You need to use `together` to symbolically calculate the addition of fractions:

```
>>> a, b, c, d = symbols('a b c d')
>>> a/b + c/d
a/b + c/d
>>> together(a/b + c/d)
(a*d + b*c)/(b*d)
```

Alternately, if you have a rational expression and want to divide the numerator by the denominator, use the `apart` function:

```
>>> apart( (x**2+x+4)/(x+2)  )
x - 1  +  6/(x + 2)
```

Exponentials and logarithms

Euler's number $e = 2.71828\ldots$ is defined one of several ways,

$$e = \lim_{n\to\infty}\left(1+\frac{1}{n}\right)^n = \lim_{\epsilon\to 0}(1+\epsilon)^{1/\epsilon} = \sum_{n=0}^{\infty}\frac{1}{n!},$$

and is denoted `E` in `SymPy`. Using `exp(x)` is equivalent to `E**x`.

The functions `log` and `ln` both compute the logarithm base e:

```
>>> log(E**3)     # same as ln(E**3)
3
```

By default, `SymPy` assumes the inputs to functions like `exp` and `log` are complex numbers, so it will not expand certain logarithmic expressions. However, indicating to `SymPy` that the inputs are positive real numbers will make the expansions work:

```
>>> x, y = symbols('x y')
>>> expand( log(x*y) )
log(x*y)
>>> a, b = symbols('a b', positive=True)
>>> expand( log(a*b) )
log(a) + log(b)
```

Polynomials

Let's define a polynomial P with roots at $x = 1$, $x = 2$, and $x = 3$:

```
>>> P = (x-1)*(x-2)*(x-3)
>>> P
(x - 1)*(x - 2)*(x - 3)
```

To see the expanded version of the polynomial, call its `expand` method:

```
>>> expand(P)
x**3 - 6*x**2 + 11*x - 6
```

When the polynomial is expressed in it's expanded form $P(x) = x^3 - 6^2 + 11x - 6$, we can't immediately identify its roots. This is why the factored form $P(x) = (x-1)(x-2)(x-3)$ is preferable. To factor a polynomial, call its `factor` method or `simplify` it:

```
>>> factor(P)
(x - 1)*(x - 2)*(x - 3)
>>> simplify(P)
(x - 1)*(x - 2)*(x - 3)
```

Recall that the roots of the polynomial $P(x)$ are defined as the solutions to the equation $P(x) = 0$. We can use the `solve` function to find the roots of the polynomial:

```
>>> roots = solve(P,x)
>>> roots
[1, 2, 3]
# let's check if P equals (x-1)(x-2)(x-3)
>>> simplify( P  -  (x-roots[0])*(x-roots[1])*(x-roots[2]) )
0
```

Equality checking

In the last example, we used the `simplify` function on the difference of two expressions to check whether they were equal. This way of checking equality works because $P = Q$ if and only if $P - Q = 0$. To know whether $P = Q$, we can calculate `simplify(P-Q)` and see if the result equals 0. This is the best way to check whether two expressions are equal in `SymPy` because it attempts all possible simplifications when comparing the expressions. Below is a list of other ways to check whether two quantities are equal, with example cases where equality fails to be detected:

```
>>> P = (x-5)*(x+5)
>>> Q = x**2 - 25
>>> P == Q                                          # fail
False
>>> P - Q == 0                                      # fail
False
>>> simplify(P - Q)                                 # works!
0
>>> sin(x)**2 + cos(x)**2  == 1                     # fail
False
>>> simplify( sin(x)**2 + cos(x)**2 - 1 )           # works!
0
```

Trigonometry

```
from sympy import sin, cos, tan, trigsimp, expand_trig
```

The trigonometric functions `sin` and `cos` take inputs in radians:

```
>>> sin(pi/6)
1/2
>>> cos(pi/6)
sqrt(3)/2
```

For angles in degrees, you need a conversion factor of $\frac{\pi}{180}$[rad/°]:

```
>>> sin(30*pi/180)                    # 30 deg = pi/6 rads
1/2
```

The inverse trigonometric functions $\sin^{-1}(x) = \arcsin(x)$ and $\cos^{-1}(x) = \arccos(x)$ are used as follows:

```
>>> asin(1/2)
pi/6
>>> acos(sqrt(3)/2)
pi/6
```

Recall that $\tan(x) = \frac{\sin(x)}{\cos(x)}$. The inverse function of $\tan(x)$ is $\tan^{-1}(x) = \arctan(x) =$ `atan(x)`

```
>>> tan(pi/6)
1/sqrt(3)                             # = ( 1/2 )/( sqrt(3)/2 )
>>> atan( 1/sqrt(3) )
pi/6
```

The function `acos` returns angles in the range $[0, \pi]$, while `asin` and `atan` return angles in the range $[-\frac{\pi}{2}, \frac{\pi}{2}]$.

Here are some trigonometric identities that SymPy knows:

```
>>> sin(x) == cos(x - pi/2)
True
>>> simplify( sin(x)*cos(y)+cos(x)*sin(y) )
sin(x + y)
>>> e = 2*sin(x)**2 + 2*cos(x)**2
>>> trigsimp(e)
2
>>> trigsimp(log(e))
log(2*sin(x)**2 + 2*cos(x)**2)
>>> trigsimp(log(e), deep=True)
log(2)
>>> simplify( sin(x)**4 - 2*cos(x)**2*sin(x)**2 + cos(x)**4 )
cos(4*x)/2 + 1/2
```

The function `trigsimp` does essentially the same job as `simplify`.

If instead of simplifying you want to expand a trig expression, you should use `expand_trig`, because the default `expand` won't touch trig functions:

```
>>> expand(sin(2*x))
sin(2*x)
>>> expand_trig(sin(2*x))    # = expand(sin(2*x), trig=True)
2*sin(x)*cos(x)
```

Complex numbers

```
>>> from sympy import I, re, im, Abs, arg, conjugate
```

Consider the quadratic equation $x^2 = -1$. There are no real solutions to this equation, but we can define an imaginary number $i = \sqrt{-1}$ (denoted `I` in `SymPy`) that satisfies this equation:

```
>>> I*I
-1
>>> solve( x**2 + 1 , x)
[I, -I]
```

The solutions are $x = i$ and $x = -i$, and indeed we can verify that $i^2 + 1 = 0$ and $(-i)^2 + 1 = 0$ since $i^2 = -1$.

The complex numbers $\mathbb{C}$ are defined as $\{a + bi \mid a, b \in \mathbb{R}\}$. Complex numbers contain a real part and an imaginary part:

```
>>> z = 4 + 3*I
>>> z
4 + 3*I
>>> re(z)
4
>>> im(z)
3
```

The *polar* representation of a complex number is $z = |z|\angle\theta = |z|e^{i\theta}$. For a complex number $z = a + bi$, the quantity $|z| = \sqrt{a^2 + b^2}$ is known as the *absolute value* of z, and θ is its *phase* or its *argument*:

```
>>> Abs(z)
5
>>> arg(z)
atan(3/4)
```

The complex conjugate of $z = a + bi$ is the number $\bar{z} = a - bi$, which has the same absolute value as z but opposite phase:

```
>>> conjugate( z )
4 - 3*I
```

Complex conjugation is important for computing the absolute value of z ($|z| = \sqrt{z\bar{z}}$) and for division by z ($\frac{1}{z} = \frac{\bar{z}}{|z|^2}$).

Euler's formula

```
>>> from sympy import expand, rewrite
```

Euler's formula shows an important relation between the exponential function e^x and the trigonometric functions $\sin(x)$ and $\cos(x)$:

$$e^{ix} = \cos x + i \sin x.$$

To obtain this result in `SymPy`, you must specify that the number x is real and also tell `expand` that you're interested in complex expansions:

```
>>> x = symbols('x', real=True)
>>> expand(exp(I*x), complex=True)
cos(x) + I*sin(x)
>>> re( exp(I*x) )
cos(x)
>>> im( exp(I*x) )
sin(x)
```

Basically, $\cos(x)$ is the real part of e^{ix}, and $\sin(x)$ is the imaginary part of e^{ix}. Whaaat? I know it's weird, but weird things are bound to happen when you input imaginary numbers to functions.

Vectors

A vector $\vec{v} \in \mathbb{R}^n$ is an n-tuple of real numbers. For example, consider a vector that has three components:

$$\vec{v} = (v_1, v_2, v_3) \in \mathbb{R}^3.$$

To specify the vector $\vec{v}$, we specify the values for its three components v_1, v_2, and v_3.

A matrix $A \in \mathbb{R}^{m \times n}$ is a rectangular array of real numbers with m rows and n columns. A vector is a special type of matrix; you can think of the vector $\vec{v} \in \mathbb{R}^n$ as a $1 \times n$ matrix. Because of this equivalence between vectors and matrices, in SymPy we use Matrix objects to represent vectors.

This is how we define vectors and compute their properties:

```
>>> u = Matrix([4,5,6])
>>> u
[4, 5, 6]                    # 3-vector
>>> u[0]                     # 0-based indexing for entries
4
>>> u.norm()                 # length of u
sqrt(77)
>>> uhat = u/u.norm()        # unit vector in same dir as u
>>> uhat
[4/sqrt(77), 5/sqrt(77), 6/sqrt(77)]
>>> uhat.norm()
1
```

Dot product

The dot product of the 3-vectors $\vec{u}$ and $\vec{v}$ can be defined two ways:

$$\vec{u} \cdot \vec{v} = \underbrace{u_x v_x + u_y v_y + u_z v_z}_{\text{algebraic def.}} = \underbrace{\|\vec{u}\| \|\vec{v}\| \cos(\varphi)}_{\text{geometric def.}} \quad \in \mathbb{R},$$

where φ is the angle between the vectors $\vec{u}$ and $\vec{v}$. In SymPy,

```
>>> u = Matrix([ 4,5,6])
>>> v = Matrix([-1,1,2])
>>> u.dot(v)
13
```

We can combine the algebraic and geometric formulas for the dot product to obtain the cosine of the angle between the vectors

$$\cos(\varphi) = \frac{\vec{u}\cdot\vec{v}}{\|\vec{u}\|\|\vec{v}\|} = \frac{u_x v_x + u_y v_y + u_z v_z}{\|\vec{u}\|\|\vec{v}\|},$$

and use the acos function to find the angle measure:

```
>>> acos(u.dot(v)/(u.norm()*v.norm())).evalf()
0.921263115666387       # in radians  =  52.76 degrees
```

Just by looking at the coordinates of the vectors $\vec{u}$ and $\vec{v}$, it's difficult to determine their relative direction. Thanks to the dot product, however, we know the angle between the vectors is 52.76°, which means they *kind of* point in the same direction. Vectors that are at an angle $\varphi = 90^\circ$ are called *orthogonal*, meaning at right angles with each other. The dot product between two vectors is negative when the angle between them is $\varphi > 90^\circ$.

The notion of the "angle between vectors" applies more generally to vectors with any number of dimensions. The dot product for n-dimensional vectors is $\vec{u}\cdot\vec{v} = \sum_{i=1}^{n} u_i v_i$. This means we can talk about "the angle between" 1000-dimensional vectors. That's pretty crazy if you think about it—there is no way we could possibly "visualize" 1000-dimensional vectors, yet given two such vectors we can tell if they point mostly in the same direction, in perpendicular directions, or mostly in opposite directions.

The dot product is a commutative operation $\vec{u}\cdot\vec{v} = \vec{v}\cdot\vec{u}$:

```
>>> u.dot(v) == v.dot(u)
True
```

Cross product

The *cross product*, denoted $\times$, takes two vectors as inputs and produces a vector as output. The cross products of individual basis elements are defined as follows:

$$\hat{\imath}\times\hat{\jmath} = \hat{k}, \qquad \hat{\jmath}\times\hat{k} = \hat{\imath}, \qquad \hat{k}\times\hat{\imath} = \hat{\jmath}.$$

The cross product is defined by the following equation:

$$\vec{u}\times\vec{v} = \left(u_y v_z - u_z v_y,\ u_z v_x - u_x v_z,\ u_x v_y - u_y v_x\right).$$

Here's how to compute the cross product of two vectors:

```
>>> u = Matrix([ 4,5,6])
>>> v = Matrix([-1,1,2])
>>> u.cross(v)
[4, -14, 9]
```

The vector $\vec{u} \times \vec{v}$ is orthogonal to both $\vec{u}$ and $\vec{v}$. The norm of the cross product $\|\vec{u} \times \vec{v}\|$ is proportional to the lengths of the vectors and the sine of the angle between them:

```
(u.cross(v).norm()/(u.norm()*v.norm())).n()
0.796366206088088     # = sin(0.921..)
```

The cross product is anticommutative, $\vec{u} \times \vec{v} = -\vec{v} \times \vec{u}$:

```
>>> u.cross(v)
[4, -14, 9]
>>> v.cross(u)
[-4, 14,-9]
```

Watch out for this, because it's a new thing. The product of two numbers a and b is commutative: $ab = ba$. The dot product of two vectors $\vec{u}$ and $\vec{v}$ is commutative: $\vec{u} \cdot \vec{v} = \vec{v} \cdot \vec{u}$. However the cross product is not commutative: $\vec{u} \times \vec{v} \neq \vec{v} \times \vec{u}$, it is anticommutative: $\vec{u} \times \vec{v} = -\vec{v} \times \vec{u}$.

Conclusion

I'll conclude with some words of warning about technological dependence. Computer technology is very powerful and everywhere around us, but we must not forget that computers are merely calculators that depend on human commands to direct them. This means it's important for you to learn how to do math "by hand" first, in order to know how to instruct computers to do the math for you.

I don't want you to use the tricks you learned in this tutorial to avoid learning how to do math and blindly rely on SymPy to do math calculations for you. That's not a good idea! I'll be very disappointed if you use SymPy to skip the "intellectual suffering" necessary to learn the new math concepts like numbers, equations, functions, etc. That's what math is all about—understand math concepts and the relationships between concepts. The part that is about rote memorized of math calculations procedures that you should have learned at school is not important at all. The tedious and repetitive math calculations is precisely what can be "outsourced" to SymPy.

To solve problems in math (or physics, chemistry, biology, etc.) the most important things are to: A) define the variables relevant for the problem, B) draw a diagram, and C) clearly set up the problem's equations in terms of the variables you defined. With these steps in place, half the work of solving the problem is already done! Computers can't help with these important, initial modelling and problem-specific tasks—only humans are good at this stuff. Once you set up the problem (A, B, C), `SymPy` can help you breeze through any subsequent calculations that might be necessary to obtain the final answer.

Most of the big math and science discoveries were made using pen and paper, which shows that scribbling on paper is a useful as a tool for thinking. With what you learned about `SymPy`, you now have access to the combination of pencil and paper for thinking and `SymPy` for calculating. It's a very powerful combination! What is a real-world problem you'd like to solve? Try modelling the problem using math equations and see what happens. Go out there and do some science!

Links

[Installation instructions for `jupyter notebook`]
`https://jupyter.readthedocs.io/en/latest/install.html`

[The official `SymPy` tutorial]
`http://docs.sympy.org/latest/tutorial/intro.html`

[A list of `SymPy` gotchas]
`http://docs.sympy.org/dev/gotchas.html`

Index

The page numbers shown in **bold** point to the concept definitions.

Notes:

No Bullshit Guide textbook series

This book is part of the **No Bullshit Guide** series of textbooks. These textbooks make the most important and useful science topics accessible to a general audience.

No Bullshit Guide to Math & Physics
Newton's laws of physics form the best-known example of the scientific approach for understanding the physical world. Starting with a few basic principles, we can build accurate models for the motion of objects and the forces that exist between them. Calculus is the language of quantitative science. Many of the fundamental laws of physics, chemistry, biology, and economics are expressed in the language of calculus. A solid grasp of this subject is essential for anyone who wants to pursue studies in science, engineering, or business. This book combines math, mechanics, and calculus topics into a single tome. By learning calculus and mechanics at the same time, readers are able to better see the connections between the two subjects, and understand concepts on a deeper level. This combined, all-in-one book is ideal for adults who want to learn (or relearn) university-level math and physics topics.

[BOOK] Ivan Savov. *No Bullshit Guide to Math & Physics*, Minireference Publishing, Fifth edition, 2014, 978-0-9920010-0-1.

No Bullshit Guide to Linear Algebra
Linear algebra is the study of vectors and linear transformations. You need to know linear algebra to understand computer graphics, statistics, machine learning, quantum mechanics, and many other areas of science, business, and technology. This book offers a comprehensive guide to all aspects of linear algebra: computational, geometric, and theoretical. A sizeable part of the book is dedicated to the applications of linear algebra, including a chapter on quantum mechanics. This book will be of interest to undergraduate students taking a linear algebra course, graduate students who need a linear algebra review, programmers, data scientists, and anyone who's interested in quantum mechanics.

[BOOK] Ivan Savov. *No Bullshit Guide to Linear Algebra*, Minireference Publishing, Second edition, 2017, ISBN 978-0-9920010-2-5.

Visit `minireference.com` to learn more about these and other titles.

Made in United States
North Haven, CT
10 November 2026

22081993R00134